Why Icebergs Float

Why Icebergs Float

Exploring Science in Everyday Life

Andrew Morris

⌂UCLPRESS

First published in 2016 by
UCL Press
University College London
Gower Street
London WC1E 6BT

Available to download free: www.ucl.ac.uk/ucl-press

Text © Andrew Morris, 2016
Images © Andrew Morris and copyright holders names in List of Figures (p.xiii–xiv), 2016

A CIP catalogue record for this book is available from The British Library.

This book is published under a Creative Commons Attribution 4.0 International license (CC BY 4.0). This license allows you to share, copy, distribute and transmit the work; to adapt the work and to make commercial use of the work providing attribution is made to the authors (but not in any way that suggests that they endorse you or your use of the work). Attribution should include the following information:

Andrew Morris, *Why Icebergs Float: Exploring Science in Everyday Life*. London, UCL Press, 2016. http://dx.doi.org/10.14324/ 111.9781911307044

Further details about CC BY licenses are available at http://creativecommons.org/licenses/

ISBN: 978–1–911307–02–0 (Hbk.)
ISBN: 978–1–911307–03–7 (Pbk.)
ISBN: 978–1–911307–04–4 (PDF)
ISBN: 978–1–911307–05–1 (epub)
ISBN: 978–1–911307–06–8 (mobi)
ISBN: 978–1–911307–68–6 (html)
DOI: 10.14324/111.9781911307044

This book is dedicated to the members of the discussion groups, whose thoughts and observations over the years have been my guide.

Preface

This is not a normal science book. It's written for people who would like to know more about science, but who may feel unsure where to start. Perhaps science passed them by at school; maybe the science shelves in the local bookshop appear daunting. Yet, despite this, many of them remain just as curious about the world around them as they were in primary school. They just lack the basic concepts needed to begin understanding things from a scientific point of view. This is a pity, and an unnecessary imbalance in their lives. After all, people manage to pick up ideas about politics and society, literature and history as they go through their lives, whether they made much of these subjects at school or not.

Why Icebergs Float aims to put this right by introducing some fundamental ideas from science in a radically different way. Unlike traditional science books it doesn't attempt to explain a particular subject, such as genetics or gravity, from a scientist's point of view. Instead, each chapter takes a question an ordinary person might be interested in as its starting point. The narrative picks up on things anyone might ask about, or may have noticed in the world about them, and discusses them before launching into relevant scientific explanation, always using plain language. It doesn't attempt to 'teach' a subject; it follows a path of inquiry you might take. In short, this is a book that starts with real life issues and leads into scientific ideas.

But how do we know the kinds of question people might wish to ask; and how they might choose to discuss them, given the opportunity? This book is based on discussions held over many years with groups of people from different walks of life, who wanted to discuss science simply because they were curious about things. The one thing they had in common was an absence of training in any branch of science, though they were generally interested in other aspects of life – politics, society, literature and history, for example – and wanted to make good the imbalance in relation to science.

In the discussion groups people asked questions they had always wanted to ask, but had felt too intimidated to do so. They exchanged their experiences and thoughts about the issue with each other, while I, as tutor, helped to identify the fundamental concepts with which the group was grappling. When more detailed knowledge was demanded, reference books, internet searches and visits to practising scientists were used to winkle out the facts. In this way an unusual and inspiring mixture of curiosity, personal experience, scientific concepts and factual knowledge was built up.

Why Icebergs Float takes a selection of these discussions as its core, with a colourful series of everyday topics – from paintings and food to illness and icebergs – acting as the entry point to a number of fundamental scientific concepts. Topics are not set out in the order you might expect in a textbook. Instead, the chapters follow the course of actual discussions, reflecting the interests of the participants and the direction they chose to take. As a consequence not all aspects of a given topic are covered, and the path taken may be unexpected. Each chapter is a discrete story; the reader can choose to read the book all the way through or dip in without losing coherence. The language assumes no prior knowledge of science, and no use is made of mathematics. If you come away from this book with a greater interest in science and enhanced confidence about tackling it, the book will have served its purpose. If you also develop an understanding of some scientific concepts, it will have exceeded it.

Enjoy!

Acknowledgements

In this book the reader is invited to eavesdrop on a series of discussions between consenting adults in a wine bar in London. Members of the discussion groups have turned out religiously, month after month, some for more than 14 years, to talk about things relating to science in their everyday lives. Ailments and travails, hobbies and holidays, offspring and ancestors, cars, computers and hairdryers and paintings are all it has taken to spark off discussion. From these, an inexhaustible torrent of questions has flowed. Less tangible aspects of life equally have inspired inquiry: how the universe arose, how humankind developed, why we behave as we do.

It is to the sturdy individuals in these groups that I owe the greatest debt: Peggy Aylett, Jane Brehony, Harry Goldstein, Debbie Karp, Carmen Kearney, Susan Kearney, Nayna Kumari, Monica Lanyado, Melissa Rosenbaum, Linda Slack, Paul Treuthardt, Penny Wesson, Emily White and Anna Wojtowicz. My thanks to them for their fascinating insights and the joyful discussions we've shared – and particularly for their permission to share some of these with a wider public. To others who have advised on the manuscript I am also indebted, in particular Sue Addinell, Richard Boohan, Peter Campbell, Dr Maralyn Druce, Dr Iroise Dumontheil, Charlotte Eatwell, Sue Jones, Daisy Minton, Ian Nash, Kate Oxley, David Oxley, Dr Kieran Quill, Professor Volker Sommer, Geoff Stanton, Derek Turner and Professor John Vorhaus.

Since my earliest days my parents, brothers and sister have encouraged an outward-looking approach to the world, savouring the colours of the passing seasons, the swirl of the departing swifts or the pumping of a steam train piston. From this a desire to understand and explain grew from the earliest age, and I am grateful for their encouragement. Equally important were my many inspiring teachers – in music, language, maths and woodwork, as well as in science. I am grateful to them all, from Mr Ewart and Mr Thomas at Bedwell Junior School in Stevenage and Mr Sills and Mr Humphreys at St Christopher School in

Letchworth to the late Dr Sandy Geddes in the Biophysics laboratory at the University of Leeds.

It's the patience and forbearance of my partner, Franco Carta, that enabled this book to be written at all amid so many other distractions. I thank him for this. Equally patient and helpful have been Chris Penfold, Jaimee Biggins and Catherine Bradley, my editors at UCL Press, who, from the very beginning of our acquaintance, have encouraged and advised me wisely.

I wish the readers well in their adventure with this book and hope a few might even be encouraged to take up the idea that inspired it – creating a discussion group and exploring the many fascinating ideas that science contributes to our lives.

Contents

List of Figures	xiii
Introduction *Getting the most from this book*	1
1. Foods We Love and Hate *The chemistry and biology of taste*	5
2. Why Old Masters Fade *Chemicals that give us colour*	16
3. Cuts are Red, Veins are Blue… *How blood delivers our vital oxygen*	26
4. The Dual Nature of Light *Particle or wave? That is the question*	31
5. Models *Developing models or finding the truth?*	50
6. How We See *From the eye to the brain*	62
7. The Brain *What it's made of, how it works*	72
8. Hormones *What they are, what they do*	86
9. Reflections on Molecules and the Body *Understanding the complexity of molecular mechanisms*	99
10. Bacteria, Viruses and Antibiotics *Why you can't take a pill for 'flu*	102

11. Floating and Density 114
    Why icebergs float

12. Tides and Gravity 121
    Holding the universe together – from Newton to Einstein

13. Energy 137
    The scientific angle on an everyday concept

14. Energy on the Move 143
    How heat energy gets around

15. Energy for Life 151
    Mitochondria and the 'three-parent baby'

16. Electricity 157
    Where does it come from, how does it work?

17. MRI and the Brain 170
    Brain scanning and what it reveals: discussion with a cognitive neuroscientist

18. Animal Culture 184
    Defining culture: discussion with an anthropologist

Epilogue 191
Reflections on what's been learned and how to take it forward

Appendix: Atoms, Elements and Molecules 195
Briefing on the building blocks of matter

Further Resources 199
Index 203

List of Figures

Fig. 1.1	Taste bud	13
Fig. 2.1	*Irises* by Vincent van Gogh	17
Fig. 2.2	Diagrams of molecules of two pigments	22
Fig. 2.3	Diagrams of molecules of two biological pigments	23
Fig. 3.1	Blood vessels surrounding the alveoli. *Image courtesy of Patrick J. Lynch, medical illustrator; C. Carl Jaffe, MD, cardiologist*	27
Fig. 4.1	Sunlight reflected off the Moon	36
Fig. 4.2	Diffuse reflection. *Image courtesy of Theresa Knott at the English language Wikipedia*	37
Fig. 4.3	An image seems to be behind a mirror	38
Fig. 4.4	The image is reversed in a mirror	39
Fig. 4.5	Light weakens as it spreads from the source	41
Fig. 4.6	Visible light as part of the whole electromagnetic spectrum. *Image courtesy of Katarina Stevanovic*	43
Fig. 4.7	Sunlight passing through the atmosphere at noon and evening. Image courtesy of Dr Rodney Schreiner	48
Fig. 5.1	Modelling a relationship. *From Wilkinson and Pickett,* The Spirit Level *(2009), courtesy of the Equality Trust, www.equalitytrust.org.uk*	52
Fig. 5.2	Two distinct models of atomic structure. *Courtesy of OpenStax College*	59
Fig. 6.1	The brain interprets what the eye sees. *Image courtesy of Edward H. Adleson*	63
Fig. 6.2	How the eye creates an image. *Copyright: University of Waikato. All rights reserved. http://sciencelearn.org.nz*	64
Fig. 6.3	Diagram of the retinal molecule (before and after a photon arrives). *Image courtesy of RicHard-59*	65
Fig. 6.4	Diagram of a nerve cell (or neuron)	68

Fig. 6.5	Microscope image of neurons in a mouse brain. *Image courtesy of Lee, W. C. A., Huang, H., Feng, G., Sanes, J. R., Brown, E. N., et al*	68
Fig. 7.1	Examples of the brain imposing form	75
Fig. 7.2	A pulse moving along a nerve cell	79
Fig. 8.1	Structures of two different steroid hormone molecules	89
Fig. 10.1	Diagram of a bacterial cell. *Image courtesy of Jynto*	104
Fig. 10.2	Microscope image of *Salmonella* bacteria (red) among human cells. *Image courtesy of the National Institutes of Health, USA*	104
Fig. 10.3	Model of a ribosome (blue) with antibiotic molecule attached (red). *Image courtesy of David Goodsell and the Structural Biology Knowledgebase*	105
Fig. 10.4	Models of two different antibiotic molecules	108
Fig. 10.5	Model of a virus. *Image courtesy of Thomas Splettstoesser of scistyle.com*	109
Fig. 11.1	Models of H_2O molecules as water and ice. *Image courtesy of P99am*	118
Fig. 12.1	Tidal bulges on Earth caused by the Moon (not to scale). *Image courtesy of Orion 8*	124
Fig. 12.2	Movement of water in ocean waves	126
Fig. 12.3	Circular motion	132
Fig. 12.4	Curvature of spacetime around a mass. *Image courtesy of NASA*	135
Fig. 15.1	The double layer of lipid molecules that makes a cell membrane	152
Fig. 16.1	The rise and fall of alternating current (AC). *Image courtesy of Sciencebuddies.org*	166
Fig. 16.2	Wire connected to the earth. *Image courtesy of Ali K*	168
Fig. 17.1	The parts of the brain. *Image courtesy of aboutmodafinil.com*	172
Fig. A.1	Models of molecules	198
Fig. A.2	Model of sodium chloride (common salt)	198

Introduction

This book is not about a particular area of science – genetics, gravity or chemical reactions, for example. It is about ideas that interest ordinary people, drawn from any area of science. The ideas come in response to questions people ask about the world around them. These questions rarely lead neatly into the traditional pattern of school subjects – physics, chemistry and biology – so the scientific material is presented in a quite different way. As there is no formal syllabus, a query about depression, for example, may lead into aspects of biochemistry and neuroscience as well as psychology and pharmacology. A structured scientific discipline helps you to learn a subject systematically by moving in carefully planned steps through an argument, but many people find the topics remote from everyday experience. This book takes the opposite tack: it focuses on issues of relevance to everyday life, sacrificing the orderly build-up of knowledge within one subject.

As a result each chapter explores a quite distinct area of science – from hormone action to tidal flows. Each is a story in itself and can be read at any point; later ones do not depend on points made in earlier ones. However, the stories have been arranged so that the content of one links to that of the next. The format of each chapter is similar, first presenting questions arising from the world around and then some of the scientific ideas that flow from them. As further questions arise the pattern repeats, forming a type of dialogue that blends science with everyday observations.

You might well wonder how the kind of questions and observations that interest ordinary people can possibly be captured by an author trained in the scientific disciplines; how can an authentic dialogue be constructed? This is a fair point and an important one. It is only too easy for a trained professional to forget how a subject looks to someone new to it: the terminology, the assumptions, the strangeness of unfamiliar ideas. In this book all the material is derived from actual discussions that have taken place with groups of people with scarcely any background in science. They meet every month in the informal setting of a wine bar to raise questions and make comments on things they have noticed that caught their interest. No topic is out of bounds and no rules constrain the path the discussion might take. The aim is to reach into scientific concepts only after questions have been raised spontaneously and people's initial thoughts and conceptualisations expressed.

The format of the stories that follow reflects this approach. Points raised by participants in discussion groups form the starting point and their thoughts and ideas are expressed first, mainly using verbatim quotes. Once the underlying scientific theme has been identified, explanation follows which itself leads into a further round of questions and observations. As the initial query is satisfied new questions, more numerous than the original ones, proliferate as fascination with the topic takes hold. At some point, after several iterations, a truce has to be called to avoid mental collapse! The stories, like the discussions upon which they are based, are brought to an end, often leaving further questions hanging in the air. Such is the fate of curiosity-driven discussions.

The approach, designed specifically for non-specialist readers, is intended to make learning more effective as well as to attract and motivate. Research on learning suggests that before trying to teach a new concept it is important to bring out the prior ways of thinking to which a person has become accustomed. For example, when children are first taught that the Earth they live on is round it conflicts with their prior understanding. As a result they are liable to feel uncomfortable and may reject the unfamiliar idea, inwardly at least. The concept of a flat Earth needs to be discussed first, to reveal the strengths and weaknesses of their deeply held conviction before persuading them to take on the strange new one with all its apparent absurdities.

Of course there are many other challenges in trying to learn about science in addition to understandable resistance to strange new ideas. It is known from research about science education that the way in which language is used can be a barrier. Unfamiliar terms and complicated, formal modes of expression may appear off-putting. If you find yourself struggling with a complicated new concept it may be best to read through it lightly, overlooking things you don't completely understand, or even to skip it and then continue where the going is easier. You can then come back to the complex section later on and see if it is more comprehensible second time round. In this book efforts have been made to use plain English; much of the text is drawn directly from the words used by people in discussion groups upon which it is based. Mathematics is not used; the emphasis throughout is on developing conceptual understanding, often through use of metaphor and visual images.

As a consequence it is important to make clear that this book does not convey a complete sense of what science is. Like any great endeavour science has many aspects: practical, mathematical, historical, as well as conceptual. By missing out on mathematical argument this book is unable to convey the true elegance of some theories. In the same way it does not give a sense of how the ideas it outlines were arrived at. It misses out on the nature of the experiments and observations that led up to the concepts we have inherited. Nor does it convey much of the human story that lies behind the historical development of scientific understanding. Fortunately, there is today a huge offering of excellent books that do cover these aspects. It is also true that the chapters do not convey a complete introduction to a given topic – they do not intend to. By following the path of actual discussions led by lay people, some aspects that may seem important to a scientist will be missed. For readers who wish to follow up with a fuller account of a topic, good, readable books are available in many areas of popular science (see Further Resources section).

With these reservations in mind, it may be helpful to set out reasonable expectations of what you, the reader, may get from the book. Its humble aim is to help you put science on an equal footing with other subjects that you are likely to have encountered in your adult life. Simply by being citizens we engage to some extent with topics in political science: we learn about voting and parliaments, governments and laws. Through our workplaces and newspapers we get some insight into economics – the way income and taxation, investment and borrowing work. Literature enters our lives through drama and novels, sociology through our engagement in communities and families, history through books and local interest groups. By these means we pick up concepts – about child-rearing, political parties, environmental issues, for example – and gain a certain amount of subject knowledge simply through living our lives. Yet this happens less easily and less often in the realm of science. Many people feel at a loss when it comes to explaining how electricity gets to their kettle or why leaves turn yellow and fall from the trees.

It is reasonable to expect that this book will help begin to set this right. People in discussion groups report on how the experience builds their confidence about scientific ideas so they feel more inclined to respond to scientific articles and broadcasts and to engage more freely in discussions. Some are surprised to see how they have sometimes been deceived by the bluster of others who purport to understand science. A few choice questions can soon remedy this. It is not likely that the book will fill you with factual knowledge that you will retain indefinitely. Life is not like that, nor is factual recall the ultimate reward of science. Most of the facts revealed in discussion groups are found after the event by searching in books and websites. The important gain from talking and reading about scientific ideas is a slowly developing repertoire of major concepts that gradually connect up. Rather like discovering a new city, you get to know a few specific areas first then gradually see how they are related.

Fundamental concepts such as the structure of molecules, the nature of gravity or the make-up of cells emerge from particular inquiries into ice or tides or 'flu. Once grasped they serve ever after in explaining a host of further phenomena – medical conditions, the strength of glues or the origin of the universe. Of course in a single book only a few of the fascinating areas of science can be opened up. This one introduces some, but leaves so many others untouched – genes and evolution, earthquakes and tectonics and the origins of *Homo sapiens*, for example. Fortunately the records of discussion upon which this book is based cover a vast range of scientific topics. With some 232 records available at the time of writing (and still growing), there's plenty of scope for many more chapters to come.

To bring the stories to life, actual words used by people in discussion groups are retained wherever possible. The names given to the characters are fictitious, and occasionally the contributions of participants have been adapted to make the reading simpler. The characterisation of individuals conveys something of the spirit of discussions and of the nature of the dialogue, but does not represent accurately any particular individual in the groups. Most of the participants are women, and they come from many walks of life. Some work in the media, some in the arts, several in charities, one or two in IT and some in the NHS and other therapeutic services. Their ages range from 25 to 75 and they come from various cultural and ethnic backgrounds. By definition they all place a high value on education in adult life and have gone out of their way to pursue it in an area they mostly missed out on earlier in their lives. I am grateful to them for all they have contributed to this book, and hope their voices help you and others enjoy the exploration of unfamiliar scientific ideas.

1
Foods We Love and Hate

Sally works in the lively office of a children's charity in central London. She had been chatting with her colleagues over lunch one day about which foods they liked and disliked. Her own pet hate was mushrooms, something she had always disliked, especially the musty old smell of the things. That evening Sally was due to meet up with her fellow enthusiasts in a science discussion group at a local wine bar. She decided to bring up this topic to see what others in the group felt and to discuss the underlying science together.

A fascinating exchange of experiences and thoughts ensued, leading to an exploration of the varied substances that flavour our food and the ways in which our bodies respond to them. Everyone seemed to have something to say about their food preferences. Dominic, a retired journalist, had disliked avocados until he was about 30, at which point he had unexpectedly developed a taste for them. Amy, a young woman in her mid-20s, had noticed that some foods you come to like as an adult, such as olives, may seem horribly bitter when you are young. Helen, who had had a lifelong aversion to cheese, had a personal theory about how this had come about: unhappy memories of having the ghastly stuff forced on her when she was a child.

This brief skirmish with likes and dislikes threw up some interesting ideas about how these preferences might have arisen – possible patterns and causes. Growing older seemed to play a part and this chimes with recent research. A lot of people do appear to become less fussy as they get older, perhaps because it can be socially awkward to reject certain foods as an adult. Another theory is that we tend to be cautious about foods that are new to us and, of course, for a young child many things are new. Talking of young children prompted Amy to comment on how unrestrained children tend to be in their demands for sweet things, often pestering their parents to the point of exhaustion for unhealthy kinds of snacks and fizzy drinks. Her suggestion was that perhaps there may have been some kind of evolutionary explanation for this. Hadn't she heard that it was in the interests of early humans to feast on sweet things as soon as they encountered them because there was no certainty about when they might next find any? Didn't sweet berries and fruits provide a highly valuable source of sugar in a time when it was scarce?

Evolution

Sugar is indeed one of the foods that yield plentiful energy when digested. In the time before agriculture, when chance played a major part in what foods you might stumble upon, sweet things would certainly have been very beneficial, though much scarcer than today. Darwin's idea of evolution through natural selection means that individuals best fitted to their environment would become more and more numerous in the population – not through any design intention, but simply by surviving longer and reproducing more. The high energy content of sweet foods may well have conferred an evolutionary advantage to those who sought them out.

So attraction to sweet things would have developed over millennia as a common human trait – a perfectly reasonable and successful feeding strategy, at a time when sugar remained in relatively short supply. For better or worse, the situation today is quite different. Sugar supplies are plentiful in highly processed forms, but we continue to express these ancient preferences, although they are no longer so advantageous for our survival – quite the reverse in many parts of the world. Evolution proceeds at a much slower pace than human cultural development. Celia summarised our modern predicament succinctly, based no doubt on personal experience: 'Isn't it odd,' she observed, 'that however full you feel after a hearty roast, you always have room for something sweet?'

Sarah wondered whether evolution might have played another role in the development of taste: signalling to the brain which kinds of food to avoid. As Amy had noted earlier, olives are not too popular with children; the love of their slightly bitter taste seems to develop later in life. Could this tendency also have evolved because bitter fruits may also be poisonous fruits? Has our sensitivity to what we perceive as bitterness evolved as a protection against accidentally eating poisons? Are poisons usually bitter in fact?

Bitterness

Fortunately there are places where tastes and smell are investigated scientifically. A study at one of them, the Monell Centre in Philadelphia, has investigated a long-held assumption among scientists that a bitter taste evolved as a defence mechanism to detect potentially harmful toxins in plants. Subjects in the study were genetically tested and then asked to rate various vegetables for taste. The evidence suggests that we are able to detect bitter toxins with our sense of taste, and genetic differences in our bitter taste receptors affect how we recognise foods containing a particular set of toxins.

This evidence bears out some of the speculative thoughts of the group: that the bitter sensation we associate with olives or broccoli appears to have developed over evolutionary time to help protect us from potential toxins. It also throws light on why individuals respond differently to a particular food, by linking taste to genetic differences. But as so often when we encounter scientific evidence, satisfying one inquiry, far from closing down a subject, seems to provoke even more questioning. If the molecules in our taste buds have evolved to detect molecules from toxic plants, what is it that gives us the actual sensation of bitterness? What is it that translates the presence of some particular chemicals in the taste buds into a subjective feeling – something that, at least when we are young, tends to turn us off such foods?

The chemistry of taste

With talk of bitterness and taste buds, the conversation turns naturally to what is actually happening in the mouth when we taste something. The idea that four principle tastes are associated with different regions of the tongue seems to be one of those pieces of information that actually stick from biology lessons at school, however long ago: salt, sweet, sour and bitter. Dominic, whose journalistic career had taken him around the world, raised the question of a fifth taste, umami. Well known in Chinese cuisine, this is associated with monosodium glutamate, an additive used to boost the meaty flavour in dishes. It occurs naturally in a wide range of foods that contain glutamate, including fish, cured meat, mushrooms and breast milk.

What those around the table did not know is what is happening chemically when each of the tastes is being experienced. It turns out that each of the five basic tastes is associated with an entirely distinct category of substance in the food. Different kinds of chemical trigger off different kinds of receptor molecule in the taste buds (explained later in the chapter). These receptors in turn send different kinds of signal to the brain. No prizes for guessing that it is sugar molecules that trigger the sensation of sweetness. No great surprise either to learn that sourness is felt when acids are detected in food. In fact the very word 'acid' derives from the Latin *acidus*, meaning sour.

More of a revelation is that the sensation of saltiness is simply the result of metal ions in food and drink hitting the palate. Ions are individual atoms, rather than the more complex molecules, which have gained a slight electric charge by losing or acquiring extra electrons in their internal structure (every electron carries a small negative charge). It is normal for the atoms in salts such as sodium chloride, for example, to separate out as ions in this way as they dissolve in water. Typical ions in food are sodium, potassium and iodine – all vital for the functioning of our nerves and other systems.

The bitter taste found in wine, beer and many vegetables corresponds to a class of chemicals called alkaloids which are present in many foods. Caffeine, nicotine, strychnine and morphine are examples of well-known alkaloids. It seems that these substances evolved as toxins in many plants precisely because they deterred herbivorous animals from eating them. Strangely this didn't seem to put us humans off coffee for long!

Finally we turn to umami, formally recognised in 1985 as a fifth distinct basic taste after it was found to be associated with a distinct receptor in the mouth. Umami is simply the taste buds' response to the naturally occurring substance glutamate, found in a very wide range of foods. It is frequently described as the 'meatiness' taste.

The psychology of taste

This introduction to the chemical basis of taste seemed interesting to people in the discussion group, if a little daunting. As often with chemistry, the number of unfamiliar terms can be off-putting. How do you get a feeling for a word such as 'alkaloid' if it has played no part in your life and doesn't connect with anything you know? The discussion took off from the basic chemistry of food and led to a more specific question: what is it in the tongue that actually picks up these various chemical sensations – the acids, the metal ions, the sugars? What are these so-called 'taste buds' we talk so loosely about?

Before a foray into the biology of the tongue could even begin, however, Helen – never one to let things hang – interrupted to bring us back to her earlier point about childhood experiences. 'Surely psychology has a crucial part to play in all this?' she rightly asked, reminding the group of her enduring memory of having been made to eat cheese as a youngster. Talk of childhood memories inspired others to chip in with recollections of a favourite children's story book. Hadn't Babar the Elephant sadly died after eating a poisonous mushroom? An interesting point for Dominic, who had spent much of his life abroad, was that the author of the Babar books was French. Given that hunting for wild fungi was a normal part of life in rural France, he wondered aloud whether the books had a hidden purpose: to warn children of the dangers of eating mushrooms indiscriminately? Talk of early years' reading reminded Amy of her lifelong antipathy to Turkish Delight, with its strange colours and wobbly texture. Could this have been linked subconsciously to its role in *The Lion, the Witch and the Wardrobe*, where the sweet is used maliciously to entice a young boy?

Our emotional responses to food do not seem to be limited to our memories of experiences in childhood. As the Turkish Delight discussion suggests, maybe other perceptions play a part too: the colour of food for example. As Helen realised, you don't see many blue foods – apart from a few kinds of berry. In fact she recalled sitting in a restaurant once under a blue light and feeling distinctly uneasy about tucking into her rice dish. Could it be that the colour blue is associated with mould, which is sometimes poisonous, she speculated?

This turn of discussion about the psychological aspects of food preferences led Julie to raise an even bigger question, particularly in relation to sugar: 'What about addiction? Is this a psychological matter or does biology play a part?' This is, as you might expect, an active area of neuroscience research. The Oregon Research Institute has used MRI scanning to conclude that sugar stimulates the same brain regions as drugs such as cocaine. Furthermore, heavy users of sugar develop tolerance (needing more and more to feel the same effect), which is a symptom of substance dependence.

It looks as though our current understanding of addiction is based on both psychological and biological research. Certain studies suggest that addiction is genetic, but environmental factors, such as being brought up by someone with an addiction, are also thought to increase the risk. As we find so often in trying to understand the science behind the way we behave, the nature versus nurture argument fails to get us very far. Our upbringing and other environmental factors play a part, but so do biological factors, including those we inherit. The balance will of course vary from individual to individual, as you might expect given the marked variation we see in our responses to sugar (or any other ingredient) in our diet.

Diabetes and hormones

Talk of sugar stirred Dominic to explain something of the nature of diabetes, a condition he had acquired later in life. In this type of diabetes (known as Type II) the level of glucose (a kind of sugar) in your blood is no longer properly regulated as a result of failings in your body's insulin system. Insulin is a naturally occurring substance manufactured in the pancreas, an organ 15 cm long that lies close to the stomach. It is a large molecule, one of many hormones in our bodies, whose particular job is to regulate the amount of glucose circulating in the blood.

Sugars are a class of substances that include fructose, lactose and sucrose (table sugar) as well as glucose. In the bonds that bind together the atoms in sugar molecules lies the energy that we need to keep our bodies ticking over. Too little glucose and our metabolism simply switches off: too much and damage ensues in organs such as the kidneys or retina, and to the cardiovascular system. For people such as Dominic, dealing with Type II diabetes means regulating the amount of sugar in their diet and taking other practical measures such as increasing physical exercise and losing weight. The other kind of diabetes, Type I, is the result of the cells that produce insulin in the pancreas actually being destroyed in an autoimmune response – that is, the body actually attacks itself. The ultimate cause of this is not known. People living with Type I have to inject insulin in order to regulate their sugar levels.

The crucial role of insulin in maintaining the right level of glucose in the blood sparked off serious discussion about the role of such chemicals in our bodies. 'OK, you've told us it's a hormone, but what does that mean?' was Helen's immediate response. By chance the answer to this question was ready to hand as the group had once organised a visit to a clinical endocrinologist at a local hospital (see chapter 8 on hormones). She had surprised the group by explaining that a hormone wasn't in fact a type of substance at all. It was simply a generic word applied to any kind of chemical in the body that is manufactured in one place but delivers its effect in a different one. So insulin is called a hormone because it is produced in the pancreas but acts in the bloodstream. Adrenalin is another example; it is produced in the adrenal glands, close to the kidneys, but acts on tissues throughout the body.

Talk of hormones and the earlier visit to an endocrinologist suddenly reminded Celia of something the endocrinologist had said. 'Hadn't she talked to us about hormones in the stomach? Wasn't she working on hormones connected with appetite?' she asked. The idea that there was more to appetite than the simple need to fill up set off new observations. 'It's not like filling up the tank of a motor car, is it?' said Sally, recalling a friend of hers who 'eats five Mars Bars a day and is still skinny. It all seems to depend on your metabolism, and this seems to change around the age of 30.' We needed to find out more about the role of hormones in stimulating appetite, and to find out why 'metabolism' differed from one individual to the next. Checking out the meaning of metabolism seemed a good starting point.

Metabolism

Metabolism describes the chemical processes that occur within a living organism in order to maintain life. It has two complementary aspects: a destructive one, in which substances that you take in are broken down to produce energy and waste, and a constructive one, in which the various chemical components of the body are built up. Metabolic rate simply means the speed at which your body burns energy. If it is fast, you burn up energy from food more quickly and remain thinner. The five Mars Bars a day person must clearly have had a high metabolic rate. Your rate is influenced by many things including the genes you inherit, your hormones, your gender, ethnicity and age – and of course we differ from one another in these respects.

The way in which hormones affect appetite is a lively area of research today. There appear to be many kinds of hormone that signal a state of either satiety (being full) or hunger to the brain. It is being discovered that some of these hormones circulate in the bloodstream and can directly influence nerve cells in a part of the brain (the hypothalamus). One hormone, called ghrelin, is secreted when the stomach is empty, signalling hunger, and ceases to be secreted when the stomach is full and stretched. It acts on brain cells, increasing the sensation of hunger and the release of gastric acid. Another hormone, leptin, has precisely the opposite effect, signalling the stomach is full. It interacts with the same cells in the brain as ghrelin.

There seem to be complementary systems in the body, one acting to promote, and one to inhibit, the same process. Like the systems that prevent sugar levels rising too high or falling too low, the body's regulation often acts like a thermostat in the home, switching on and off at the right moment to maintain steady conditions. The possibility that drugs might be designed to act directly on the hormones that regulate appetite is, as you might expect, an active area of research into obesity. By inhibiting the ghrelin hormone, appetite has been shown to be reduced. This leads in turn to lower food intake and reduced body weight, giving rise to speculation that this might be a possible way to treat or prevent obesity.

This foray into the chemistry of appetite is fascinating. It gives a sense of how the chemicals that make up the food we eat interact with chemicals in the body, which correspond in some way with how we feel – hungry, satisfied, craving more or lacking appetite. The scientific detail revealed may be too much to take in at first, but the general point – that the physical process of eating and the emotional reality of hunger are linked through the interaction of known substances – is illuminating. Even more inspiring is the fact that the nature of these interactions is under investigation and is already partly understood: body and mind linked in one continuous interplay.

But, as Celia suggested after this surfeit of chemistry: 'Can't we get back to the original point: what is it that gives us the sense of taste, and why do we all react differently to the experience of it?' One thing that everyone round the table seemed familiar with was the concept of taste buds. We all use the phrase and know they are to be found on the tongue. We may even know they are capable of distinguishing the four or five basic tastes: sweet, sour, acid, bitter and umami. The inevitable next question was posed by Julie: 'What actually are "buds" and how do they work?'

Taste buds

The idea of taste buds is something we may remember vaguely from school without being able to recall what they actually are. My human biology textbook describes them as 'tiny structures on the surface of the tongue and elsewhere in the mouth'. It goes on to explain that within the buds are cells of a particular kind, known as receptor cells. Cells of this class abound all over the body; their role is to respond to signals by binding to the signal molecules that are passing by outside the cell. Special molecules called receptor molecules, found on the external surface of the cells, in effect 'receive' these signals. So there's a hierarchy: on the tongue are the taste buds; these are made up of cells; and on the cells are receptor molecules which pick up the molecules from the food.

The key to the orderly regulation of the body's physiology is that each type of receptor recognises only one specific type of molecule. In the case of taste buds, the receptor molecules are located on tiny 'taste hairs' that form part of the receptor cells. These are the molecules that do the business: they respond uniquely to one type of food molecule or another. Chemicals in the food dissolve in saliva and attach to these receptor molecules. Each receptor responds to a particular flavour ingredient. Sugars set off the sweet taste, acids set off the sour, metal ions (such as sodium) activate the salty and alkaloids (like caffeine) the bitter.

That's how specific tastes are initially detected, but how are they experienced in the mind? The link between the physical act of tasting and our affective response is down to some important nerve fibres that are wrapped around the cells that contain the receptors. When the receptor molecules in the taste buds detect some specific ingredient in the food we eat, these specific nerve endings get activated. This activation relays a corresponding message to the brain, which gives us our perception of the taste. In the diagram (Fig. 1.1), the taste pore is where food molecules enter the bud and the afferent nerve is where the electrical signal leaves for the brain.

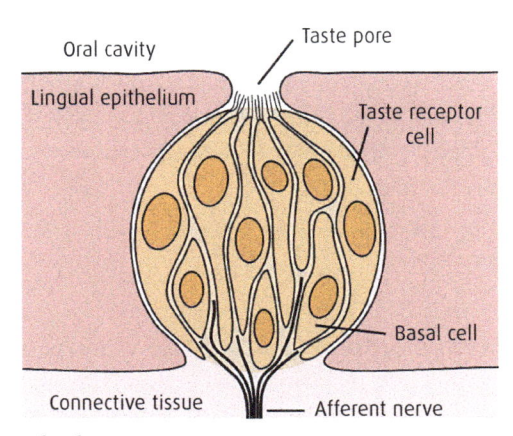

Fig. 1.1 Taste bud

Smell

Having explored something of the actual mechanism of tasting and gained some insight into the molecules that cause the sensation of sweetness or bitterness, our discussion reverted to things we might notice in everyday life. A key point introduced by Amy was that surely our sense of the flavour of food cannot be entirely down to our taste buds: 'after all, when your nose is blocked, you lose an awful lot of the pleasure of eating tasty foods.' Smelling food is as important as tasting it – a point wine experts remind us of when they recommend swilling the drink around in a broad rimmed glass before judging its quality. When we swirl wine around to smell the aroma, volatile chemicals in the wine evaporate, enter our noses and send signals to the brain, which it interprets as flavour. Most, then, of what we commonly perceive as taste is actually provided by our sense of smell.

Mention of the sense of smell launched a further stream of thoughts and observations in the group. Celia wondered about the enhanced sense of smell that so many animals seem to have. Some breeds of dog can smell truffles buried underground, others detect illegal drugs and yet others can sense a fox miles away. Someone had heard that even the smell of an impending snowstorm can be detected by some animals. The underlying mechanism of smell shares many of the features of the system of taste. Molecules emanating from an odorous source are detected by receptors in the cells of the olfactory system of the nose. As with the mechanism of taste, these interactions then trigger off nerve signals that inform the brain. An interesting observation made by Dominic, who had travelled in Asia, was that some foods seem to give off an appalling odour yet, if you can manage to get through them, have a gorgeous taste. He talked about a fruit called durian that grows in Southeast Asia and is regarded by some as the king of fruits, by others as simply revolting. Celia brought this paradox closer to home, recalling how foul-smelling many popular cheeses are, while still being renowned for their outstanding flavour.

Conclusion

Talk of cheese and wine, fruit and fungi brings this exploration of taste-inspired science full circle, back to its original starting point: the variety of flavours in food and our personal likes and dislikes. As with many science discussions, people's observations and experiences of everyday life have led to an exploration of many important scientific concepts. We have seen that food is made up of a huge variety of different substances, from sugars and proteins to acids and salts. We have seen how structures in the tongue and nose are able to detect and differentiate between these various substances, and how each of these can trigger a chain of events culminating in nerve messages to the brain – which, in turn, give rise to our emotional reactions.

These are fundamental concepts that we encounter over and over again in exploring the body – the structure of molecules, the nature of receptors, the linkage between sensory cells and nerve cells. All are seen to recur when we consider the processes of seeing, hearing, digesting, lifting and countless other bodily processes. From a simple inquiry about the foods we like and dislike, fundamental scientific concepts are soon encountered. We may not have actually come up with an answer to the original question – why we like some foods but not others – but, like so many lively discussions based on a casual observation from everyday life, it has been an interesting excursion.

The discussion about the taste of food started innocently enough with everyone in the group having something valuable to say about things they had liked and disliked. It also reminded people of some of their early experiences, whether memories of the stories they had once read or the indignities they had suffered with food they disliked. Yet, soon enough, discussion had reached far into some powerful concepts in science – in particular the idea that all substances are ultimately made up of molecules of various sorts. What is more, the grouping of chemicals into classes, such as salts, acids and alkaloids, became clear, as did the fundamental concept that molecules are broken down and rebuilt during the course of a chemical reaction. The energy stored in the bonds that link the atoms in each molecule were seen as an ultimate source of energy for life. The universal nature of fundamental concepts in chemistry was reinforced when they appeared one day in another, quite distinct discussion, with a different group of people. This discussion had been sparked off after a member had visited an art museum where she had read about efforts to restore the colour in some post-Impressionist paintings. It forms the starting point for the next chapter.

2
Why Old Masters Fade

The Van Gogh Museum in Amsterdam must surely count as one of the most colourful art museums in the world. The eye is assailed from all directions by the richest of tones, often heightened by the juxtaposition of complementary colours. The blue flowers against a yellow background in Van Gogh's painting *Irises* is a classic example of this dramatic type of contrast (Fig. 2.1).

Julie described the excitement of the paintings in a discussion session shortly after returning from a break in the Netherlands. She went on to raise an interesting question. An information display in the gallery had pointed out that some of the colours in the paintings were not in fact as the artist had originally laid them down. Some had faded, but others had not. As a result the hues and contrasts did not always reflect the artist's intentions. 'Why', she asked, 'do colours fade?' In discussing the possibilities, Sarah remembered hearing that some colours fade more rapidly than others, so distorting the overall balance. Jean, who had been to an exhibition on colour at the National Gallery in London, recalled that even within a single class of colours some particular ones fade faster than others. She had seen a painting of the Marquise de Seignelay by Pierre Mignard in which the sky had faded from blue to grey, whereas the rich velvet of the gown remained a brilliant azure.

What was the difference? What is it in paint that actually gives the colour? Is the cause of colour in paint any different from that of anything else – leaves, flowers, clothing or skin?

'Pigment, pigment, it's the pigment! I've just remembered, you get it in paint and in skin,' exclaimed Sonya, recalling a rarely used word. 'Yes, it's pigment all right, that's what artists used to mix with egg yolk or oil to make paint, wasn't it? But what the heck is pigment? Is it a chemical? Something natural in the body or something you make?' At this point those who had seen the National Gallery exhibition found

Fig. 2.1 *Irises* by Vincent van Gogh

themselves at an advantage as they had actually seen pigments: dozens had been on display in small glass jars and cabinets. Many looked similar in form – a brightly coloured heap of finely ground powder: vermilion, azure, ochre. But on closer inspection they turned out to be of two quite different kinds – almost literally as different as chalk and cheese. One group was essentially minerals, hewn from the ground in diverse parts of the world, while the others had been concocted from various living things.

Pigments

The most precious blue pigment, the one used to dress the Madonna in paintings, for example, is made by grinding up lapis lazuli, a rare kind of rock found in Afghanistan. Less expensive to produce and to transport was azurite, a different type of blue; this is made from ground-up copper carbonate found more commonly in parts of Europe. Unfortunately the cheaper one fades over the centuries, whereas the expensive one holds its intensity indefinitely – hence the difference between the faded sky and brilliant gown in the Mignard painting.

Other pigments made from ground-up rocks include the dark yellow ochre (from the mineral limonite) and the rusty red of iron oxide.

For sheer brilliance in the red zone, however, no mineral can match the splendour of cochineal, the colouring used in upmarket patisserie as well historic paintings. But cochineal is not in fact derived from a mineral at all; it is an example of an entirely different source of pigment – organic matter. It comes from the ground-up outer shell of a particular kind of tiny beetle found in South America. The beetle's body is filled with a red substance that holds its own when mixed with oil or egg yolk, providing the brilliance that we still see in the garments of the wealthy depicted in Old Master paintings. Other kinds of life form supply the many kinds of green, deep red and yellow that we associate with traditionally dyed wools and cottons: the pigments in so-called organic dyes. Indigo, for example, is extracted from the leaves of tropical plants, sepia from the ink sac of cuttlefish, and madder from the roots of plants in the coffee family.

So now we have an idea of the origin of colour at one level. Pigments, which are chemicals in powder form, made from a variety of substances, give colour to materials. In the case of Old Master paintings, they are combined with some kind of glue, such as egg yolk, or oil to make them more adhesive. With these basic ideas about pigments established, a deeper set of questions sprang from the group. What are pigments actually made of? Why are they coloured? What is actually happening when we see a colour, as opposed to something colourless or plain white or black?

Why things are coloured

These questions led into a very interesting area of chemistry: the nature of the molecules that give rise to the sensation of colour. Pigments are, of course, a kind of chemical, just as washing-up liquid, proteins and aspirin pills are, and all chemicals consist of vast numbers of tiny structures – molecules. At their miniscule level, molecules come in all shapes and sizes: they may be small, large, flat, globular, symmetrical or knobbly. However, they are all composed of atoms of various kinds bonded together (see Appendix for explanation of atoms and molecules). As the nineteenth century progressed chemists gradually developed the technology to analyse what the chemicals were in any given pigment and, later on, to find out what the molecules of each pigment actually looked like. In this way it could be worked out which atoms were in each molecule and, crucially, how they were arranged in the overall structure.

It turns out that there are patterns in the chemical composition of pigments. Many of the minerals that were ground up to make pigments were found to contain one of a particular group of metal atoms; examples of these are cadmium, chromium, cobalt copper, iron and lead. The names of these metals are still reflected in artists' colour charts: cadmium red, chrome yellow and lead white, for example. Known as the 'transition metals', these atoms share a particular property that gives rise to colour in pigments. Like all atoms, they are composed of a tiny core known as the nucleus and a number of tiny, electrically charged particles called electrons that exist around the nucleus. These electrons are able to exist only in certain discrete states of energy – they cannot exist with an amount of energy in between these allowable levels. This rather bizarre fact is hard to visualise as most things we are aware of seem able to exist with any amount of energy we choose (within practical limits) – cars can go at any speed, radiators can be adjusted to any temperature.

This limitation on electrons (or any other particle) – that they cannot exist with any amount of energy, only with specific amounts – is one of the great counterintuitive concepts of the radical 'quantum theory' developed in the early twentieth century. The situation can be likened to books on a bookshelf. They can only exist on one shelf or another, not in positions between the shelves. It was soon realised that you can understand the origin of colour by imagining the electrons inside the atoms shifting from one state of energy to another. Light impinging on an atom can shift an electron from one level of energy to another. The light has to carry just the right amount of energy to enable an electron to raise its energy from one level to another – no more, no less. So, in effect, light falling on the atom gives up some of its energy to the atom. This process is in fact the explanation, at the very small scale, of how any kind of light gets absorbed by any kind of substance. Energy contained in the particles of light (known as 'photons') is transferred to the electrons inside the atoms of the absorbing substance. A photon carrying the right amount of energy raises the energy level of an electron and in so doing loses some of its own energy, a process we call absorption.

What is most intriguing is that the atoms of any substance can only absorb a particle of light that has exactly the right amount of energy to enable an electron to change from one level of energy to another. Particles with the wrong amount of energy are simply not absorbed. It is like saying that only the exact amount of energy needed to lift a book from one shelf to another can be absorbed. This is the essential meaning of the word 'quantum' (from Latin for 'how much'). The implications of this for our colourful world are profound. Only particles of light of a particular energy get absorbed by a substance; and different colours of light have different levels of energy. So only some colours in the light falling on an object get absorbed; the rest is reflected away. This, the reflected light, is what we see. At this point, Julie, who had been interested primarily in Old Master paintings, understandably began to get a bit edgy. 'Here we are talking deep atomic theory when all I asked is why colours fade on old paintings. Give me a break,' she chastised gently.

I think there is a profound issue here. From countless discussions about science I have led, it seems that questions from everyday life, which can be stated quite simply, very often lead into rather advanced concepts and unpredictable encounters with deep theory. This may be why science learned at school may appear to be rather unrelated to people's actual lives. Scientific knowledge is normally acquired step-by-step, starting with simplified models of how things are – simple chemicals, simple organisms, simple pendulums. Only when you reach more advanced levels does science get closer to more realistic situations. As a result, we don't see much quantum physics or colour chemistry in GCSE syllabuses. That's why there is not much chance to talk about paintings and pigments at school level. This might be one of the many reasons people switch off from science, perceiving it as remote and dull, failing to connect with their experiences in life. As adults trying afresh to get to grips with basic ideas in science, this is the price you pay for linking science more closely to actual experience. We start with real questions and these demand deeper explanations – but it's well worth the effort.

So, renewed by this exhortation to keep at it, let's see how the absorption of light by atoms helps us address our questions. The point is, as you may have picked up at school, ordinary daylight, known as 'white light', is in fact composed of all the colours in the rainbow. In fact it is this kind of light from the Sun that gives rise to the colours of a rainbow. Isaac Newton famously showed this when he arranged a glass prism to pick up a beam of light from a slit in the shutters of his window. It produced a lovely spectrum of colours on a nearby wall.

The daylight that surrounds us ordinarily is in fact a mixture of light of all colours. So when a red post box is bathed in sunlight, why does it look red? Of course if the daylight wasn't there, for example at night time, the post box would not look red at all, but black. We only see a post box when light is shining on it: it isn't beaming out light like a lighthouse. The box is not a *source* of light; it is only visible because it reflects the light falling on it. So why, you might well ask, does it appear red if the daylight is white? Suddenly we see the picture – the red paint on a post box must be absorbing all the colours in the daylight *apart from* the red. As a result the red part of the light is the only colour reflected into our eyes. What we see is the light left over after the pigment in the paint has subtracted all the other colours: green, blue, yellow and so on. The atoms or molecules in the pigment have that special property – the electrons inside them are arranged in just the right way to absorb all these blue-green colours, leaving the remaining red to reflect off and reach our eyes.

So that is what pigments are: substances whose atoms absorb particular colours of light, leaving the remaining ones to reach our eyes. In the case of chlorophyll, the substance that gives foliage its green colour, blue and red light are strongly absorbed, leaving the green and some yellow to be reflected. For carotenoids, the substances that give carrots their colour, absorption is mainly in the blue and green zones, leaving the yellow, orange and red to be reflected. This gives the carrot its characteristic colouring.

This story of the origin of colour proved fascinating to the discussion group. At last they had some kind of explanation of what colour is and why different substances have different colours. It's because atoms in pigments soak up some of the colours in daylight, leaving the remaining colours to reflect off and enter our eyes. Yet one point still felt unclear. Chlorophyll may be green and cochineal beetles red, but surely these biological substances can't act just like the minerals discussed above. After all, leaves and beetles don't look or feel like stone. Are biological pigments really made of the same metal compounds that make up lapis lazuli and ochre?

Since the early nineteenth century biological pigments have gradually been isolated by chemists and studied in laboratories. It turned out that they were indeed quite different from the mineral ones. They were not made up from single atoms of metals such as iron and cadmium. Instead they turned out to contain rather complex molecules, each of which had something in common. Here are diagrams of two of these molecules (Fig. 2.2). The lines represent the bonds that link atoms within the molecules together. For clarity not all the atoms are explicitly drawn in, but are implied wherever lines meet. Instead they turn out to contain rather complex molecules, each of which has something in common.

Even without fully understanding what these diagrams represent, a common feature is immediately apparent: each involves an alternating

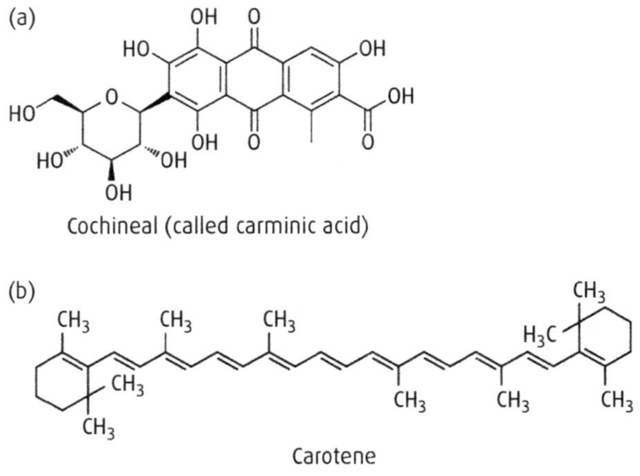

Fig. 2.2 Diagrams of molecules of two pigments

sequence of parallel lines separated by single lines. These represent respectively 'double bonds' and 'single bonds', particular ways in which electrons are shared between adjacent atoms. The details of how this pattern of bonds arises need not occupy us here, but the effect of them is to create states of energy that enable light of specific colours to be absorbed. The consequence is that electrons in these molecules are able to shift their levels of energy, just as they did in the iron and cadmium atoms in mineral pigments. In just the same way, this means that light of certain specific colours gets absorbed by the molecules, but light of other colours does not. In the same way, light of the colours that are not absorbed gets reflected back to our eyes, giving us the impression of a coloured object.

The original question that launched this excursion into pigment chemistry had been inspired by a casual observation on a visit to an art gallery. The journey that led from discussing paintings into the science of the atoms and molecules of pigments might have seemed enough to conclude the discussion, but instead it inspired further questions about colour. 'Now we have some idea about pigments in paint, what about the colours in other, naturally occurring things, like leaves or blood?' asked Mary, seeing how the idea of molecules and absorption of light might extend beyond paintings.

Well, as you might have guessed, these substances also contain pigments, and the task of identifying them has been one of the great achievements in analytical chemistry. As we have seen, the green colour of plant material is due to the chlorophyll molecule, whose job it is to

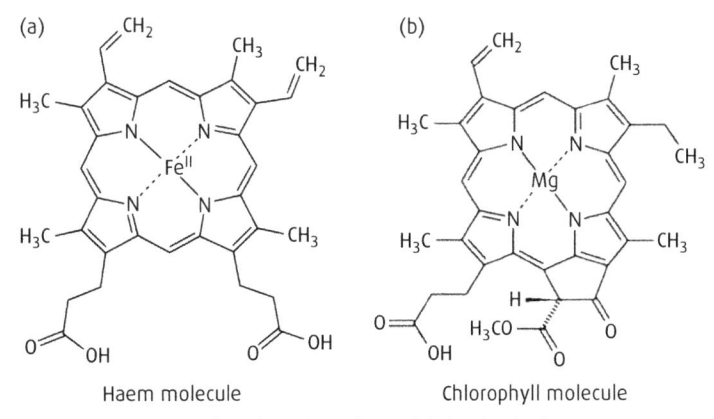

Fig. 2.3 Diagrams of molecules of two biological pigments

absorb as much light energy as possible from the Sun. The red colour of blood is due to the presence of another special molecule called 'haem'; it lies inside the haemoglobin molecule that carries oxygen round the body. A quick glance at the structure of these two molecules once again reveals the characteristic pattern of alternating double and single bonds that we met earlier, though in this case they are arranged in a quite different geometry.

One look at these fiendish diagrams was enough to conclude this part of the discussion. As one person expressed it: 'This has turned out to be complicated – molecules, double bonds, energy levels. Can't we get back to our original question: why is it that Old Master paintings fade?' What have molecules, bonds and energy levels got to do with it?

Let's pull together what we have found out so far. We have seen that traditional paint is made by mixing coloured substances called pigments with some kind of glue that binds them to the canvas. The pigment is made of molecules that absorb many of the colours out of the light that is falling on them. However, some colours are not absorbed; these ones are reflected off the paint and are picked up by our eyes. It is this unabsorbed, reflected light that gives the paint its colour. Which particular colours of light get absorbed by a molecule is determined by the energy levels that are available for the electrons in the molecules of the pigment. So the answer to the original question about the fading of colour in painting (and in fabric) is now fairly simple. It means, as you might guess, that the molecules of pigment in the paint are gradually lost over time. It's not that they somehow disappear or evaporate; they simply get altered. The bonds that hold together the atoms of the pigment molecule can eventually break if light of sufficiently high energy impinges on them, a process known as photolysis (from two Greek words, *photo*, meaning light, and *lysis*, meaning loosening).

Ultraviolet radiation is a high energy form of light and, as sunbathers are only too aware, is present in ordinary daylight. Over time, ultraviolet radiation can gradually break up the molecules in pigment, leaving smaller, colourless molecules as products. There is also a second process of degradation in which pigment molecules may react chemically with oxygen molecules in the atmosphere, a process known as oxidation. As with photolysis, this alters the structure of the molecule and, as a consequence, changes the manner in which it absorbs light of various colours. As the amount of coloured pigment in a given area on a canvas gradually diminishes, so the colour seems to us, as onlookers, to fade. So a given pigment doesn't actually change colour; it simply becomes more dilute in the mix of pigments over time.

Sometimes the colour we perceive is something of an illusion, created by a very intimate mixing of pigments. The purple coloration of a royal robe, for example, may turn out under the microscope to result from a densely packed mixture of tiny dots of blue and red pigment. If the blue and red pigments fade at different rates, the apparent colour will change. A striking illustration of the effects of fading has been demonstrated recently at the Art Institute of Chicago. A painting by Renoir, *Madame Léon Clapisson*, has been digitally restored after identifying the nature of the red pigment used. It was a highly light-sensitive molecule called anthroquinone, derived from the cochineal beetle. Over the years the strong red colour disappeared from exposed parts of the painting, and the digital restoration reveals what Renoir had originally intended. The remarkable result can be seen on the websites of the Art Institute of Chicago or on Chemistry World.

The fundamental scientific concept emerging from this exploration of paint, blood and leaves relates to how light interacts with materials. Substances, whether animal, vegetable or mineral, natural or artificial, are made of zillions of tiny molecules; these absorb light of differing colours thanks to the arrangement of electrons inside the molecules. This intricate interplay of light with a select group of molecules gives rise to the extraordinary world of colour that we humans are privileged to perceive.

Colour is a recurrent theme in discussion groups. It is, of course, a striking part of the environment in which we live and figures prominently in our interpretation of the visual world and in our emotional responses to it. We first encountered the topic in connection with our reactions to food; it reappeared when we investigated the fading of old paintings. These starting points led us to explore the workings of human taste buds and the nature of the pigments used by painters over the centuries. In the following chapter a quite different starting point, a small cut to the knee, led unsurprisingly to discussion about blood and its role in the body. Once again the path led to exploring the nature of the molecules that compose the countless substances of which all things, living and inert, are made: on this occasion the life-giving substances that supply our vital organs.

3
Cuts are Red, Veins are Blue...

'My boy came in yesterday with blood dripping from a cut on his knee,' said Mary, a mother of two, at the start of a discussion one day. Apart from her immediate concern for her poor son, she was struck later by how just how bright the red colour of blood is. She went on to recall being told anecdotally that this was only due to the blood being exposed to the oxygen in the air, not because it was actually bright red inside the body. 'Is this true?' she asked. 'What colour is blood in the body? On the diagrams we see in hospital charts, it is only red in one half of the body and blue in the other half. Does it change colour in the middle?'

It turns out that blood is in fact red everywhere in the body; it's just the shade that changes. The blue colour is just for the wall charts. The blood that comes fresh from the lungs is bright red because it has just been replenished with oxygen, but the blood that returns from the rest of the body back to the lungs is dark red because some of its oxygen has been given up. That of course is the whole purpose of the respiratory system: to get oxygen from the air into the tissues where it is needed. It's not really a big surprise to hear that blood is always red, never blue; after all, blood looks red everywhere when you see an operation in a film or on television.

With the blue blood myth sorted, discussion moved on to deeper questions: why does oxygen make blood red-coloured anyway? How does oxygen actually get into the blood in the first place? Exploring these questions led us into interesting aspects of chemistry and biology.

Oxygen is one of the gases that make up the air that surrounds us; it makes up about 20 per cent of it at sea level (a bit less up a high mountain). Essentially oxygen, like any gas, consists of zillions of tiny molecules buzzing around at very high speeds. It's hard to imagine this because we don't see or feel these molecules, but we know about them from their effects. When we breathe in, an enormous number of these molecules pass through the narrowest tubes in our lung tissue until they reach the end point of the tubes – tiny hollow cavities called alveoli. These are the ball-like shapes in the diagram (Fig. 3.1).

Here the molecules in the air are surrounded by the walls of the alveoli. Metaphorical words such as 'wall' can easily mislead us, however. We think of a wall as a robust impenetrable thing, made of bricks and sealed with mortar, but a biological wall is quite different. Yes, its primary role is to separate one space from another like its brick-built counterpart, but at the same time certain molecules are able to pass right through it; biological walls are flexible and, to some extent, porous. The walls of the tiny alveoli inside the lungs, for example, are not simply a continuous sheet of material, but are made up of thousands of cells packed one next to the other. The tiny molecules of oxygen are able to pass through the 'walls' of these cells and the spaces that lie between them. These semi-permeable 'walls' surrounding cells are known as membranes.

In fact, there is more than one way in which molecules of oxygen can get through the alveoli. They can simply diffuse passively through the gaps between the cells of the alveoli walls, or they may be actively picked up and carried through by other larger molecules embedded in

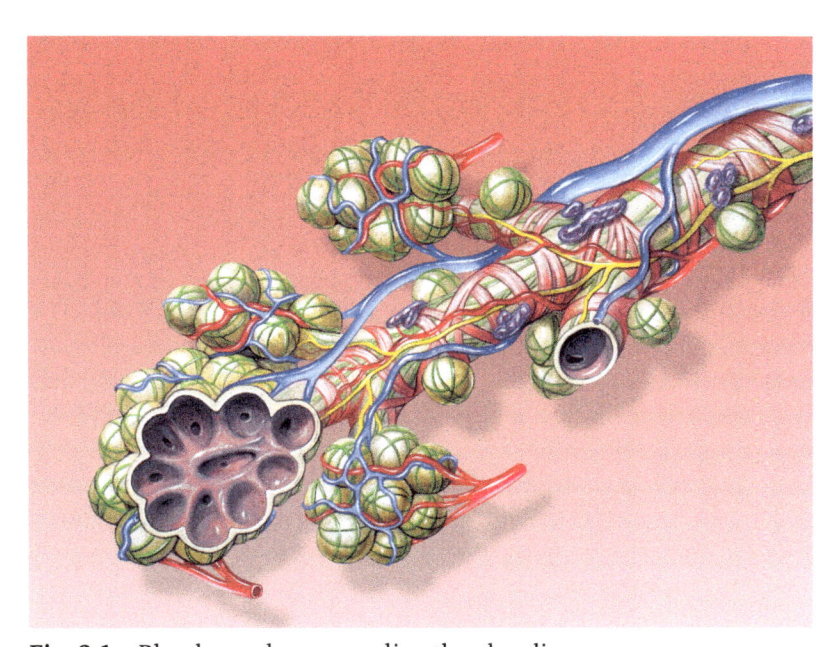

Fig. 3.1 Blood vessels surrounding the alveoli

the membranes of the cells in the wall. Sitting conveniently by, on the other side of the walls of the alveoli, are the very tiniest of blood vessels (capillaries), and the oxygen molecules are able to pass through the walls of these blood vessels just as they did when leaving the lung tissue. This enables the oxygen molecules to get into the bloodstream. As the word implies, this stream is constantly flowing so the molecules are transported on a relatively short journey to the heart. The heart, with its four chambers, not only receives the blood rich with oxygen molecules, but also despatches it directly to all the distant parts of the body.

'Yes, yes, we get all that.' A certain impatience entered the conversation. 'We do know a bit about the heart. Can we get back to the original question: what is it that makes the blood red? We thought oxygen was colourless – look at the air after all, it isn't red!' So the oxygen molecules have got into the blood, but we need to think about what blood actually is. When we see the results of a blood test it seems to be quite a complicated mixture of things. There's plasma, white cells, red cells and plenty more. The main component is the so-called 'red cells', which really do look red under a microscope. Their main job is to store a mass of huge molecules of a type of protein called haemoglobin. This name crops up regularly if you have medical tests because your haemoglobin count is an important indicator of health. It's the haemoglobin molecules that give blood its red colour.

'OK, so now we know that blood is red because it contains red cells, and these cells are red because they contain haemoglobin molecules. But we still don't know what makes the haemoglobin molecules themselves look red. Are there some kind of "redness atoms"?' No, atoms aren't 'coloured' in that sense. As we saw in chapter 2, the colour we see from the atoms and molecules in a substance is simply the colour of the light that is reflected off them, after they have absorbed various component colours in whatever light is shining on them. Different molecules absorb different amounts of each colour. Haemoglobin is a molecule that absorbs light of many colours, but not red. It is a large and complicated molecule, like most proteins. However, inserted inside haemoglobin molecules are smaller flat molecules known as 'haem', and right in the middle of each of these lies a single atom of iron (see Fig. 2.3). It is easy to imagine that this iron atom might account for the red coloration of our blood, in the same way that iron oxide gives rust its colour. This is not so, however; the red coloration actually comes from the haem molecule, not from the iron. When oxygen is present, this changes shape slightly, altering the way in which it reflects light, so making the colour a brighter red. Our whole respiration system depends on us having enough iron atoms to fill the haemoglobin molecules in our red cells and to carry oxygen molecules from the lungs to the various tissues of the body. So eat up your spinach and avoid anaemia!

As I had expected, this attempt to describe why blood is red and the role oxygen plays proved just a starting point for discussion. The original issue had been whether blood is really red or blue or both. By exploring the way the lungs take in oxygen molecules and pass them through to the bloodstream we had gained insight into some basic anatomy and cell biology. The 'red or blue' question had then catapulted us into some quite complicated chemistry – the nature of the haemoglobin molecule that picks up the oxygen molecule and carries it away to where it is needed. The whole sequence turned out to be a story of molecules:

- The oxygen molecule that abounds in air, made up of two oxygen atoms bound together
- The haem molecule, with its single atom of iron at its centre, that picks up the oxygen molecule
- The haemoglobin molecule, which transports the oxygen to where it is needed. The haem molecule is embedded in the large globin molecule.

So, despite the tempting idea that the iron atoms in blood are responsible for its characteristic red colour, it is in fact the haem molecule, the flat ring of atoms, that provides the red colour. And when an oxygen molecule binds to the iron atom, it brightens the red colour of the haem molecule.

With the cause of the redness sorted, a new question immediately arises. If blood is red, why are the blood vessels we see in our arms and legs blue? We know there is no such thing as blue blood – it's either bright red (when it has plenty of oxygen) or dark red (when it has less). Interestingly, research has been carried out into this precise question in Canada, which suggests that the colour we attribute to the veins is not the actual colour of the veins. The blue comes from the daylight impinging on our skin, penetrating through to the veins and then reflecting off the walls of the blood vessels, back to our eyes. As daylight falls on the skin and passes through it, the red part of the light gets more strongly absorbed than the blue. As a result, more of the blue component of light gets reflected back to our eyes, causing us to see the vein as blue.

This explanation reminds us of a very fundamental point about light that we so easily overlook. When we say that we 'see' something, it simply means that our eyes are receiving the light reflected off it (or emitted by it, if it is a source of light). This brings us to a fundamental philosophical point touched on in chapter 2: things don't actually have an intrinsic colour – they just absorb some colours and reflect others. If we want to press the philosophical point even further, can we even say that an object has any colour at all when it is in a dark room? After all, no light is falling on it, so it is neither absorbing nor reflecting. Is colour simply an artefact of our eyes and brains? Could a world without sighted animals be said to have colour at all? But that's surely enough questioning for one day! A simple cut to the knee has led us to the role of oxygen, the nature of haemoglobin and ultimately to the deeper meaning of colour. Enough!

Blue food, red blood, purple pigment – colour is a constant source of fascination. The changing shades of autumn leaves or the gendered colours of children's clothes – whatever the original cue, colour turns out to be a terrific starting point for exploring scientific ideas. It can lead straight into aesthetics and the psychology of perception: danger, envy and cowardice. Or it can go deep into the chemistry and biology of life, such as the shape of blood molecules or the evolution of taste. In the next chapter, an innocuous question about colour in the kitchen launches us on a journey into the nature of a very familiar phenomenon that surrounds us every day: light.

4
The Dual Nature of Light

The idea of light has fascinated people from time immemorial. In ancient Greece Empedocles likened the eye to a lantern, imagining fire to exist within it. Euclid later introduced the idea of light rays as straight lines that were for him no mere geometrical fiction. It was these rays that actually caused us to see things when they emanated from our eyes and fell on objects. I recall vividly as a child being mystified by what light was and why people seemed to bother even to talk about it. There was the stuff we played around in all day long, which then disappeared inconveniently in the evening. In winter time there were also lights you had to switch on. That was it, why the fuss? But as I grew older I remember hearing adults talk reverentially about the quality of light, declaring that 'it's so wonderful' in the Mediterranean, in the Greek islands and Spain. As an adult myself, I remember reading about the Impressionist painters making a philosophy of light, emphasising the play of natural light. Van Gogh travelled to Arles specifically for the direct sunlight that was lacking in Paris.

In secondary school the subject seemed even more remote and increasingly baffling. I realised it was important because it gave its name to a whole chapter of the textbook and a great chunk of our physics lessons. But I found it difficult to see what fiddling around with little rectangular mirrors and pins stuck in cork had to do with the sun-drenched South. Then there were intriguing things you could do with your friends, when the teacher was occupied elsewhere: navigating your way across a room with a prism to your eye, for instance, or focusing the Sun's rays on to some unfortunate wrist with a magnifying lens.

Little did I realise at that age just how rich and fascinating the concept of light would become as I grew up. The aesthetic appeal of the changing light at dawn, noon and sunset would intensify the joy of the outdoors and open up the atmospheric quality of paintings. Equally, in science, gaining insight into the nature of this evanescent, immaterial quality became a source of fascination for me – as it had proved to be for countless investigators, theoretical and experimental alike. The to-and-fro of competing theories divided great scientists throughout the centuries, and culminated in one of the most profound philosophical advances of modern times: the acceptance of inherent duality in our understanding of the world, an end to the certainties of the nineteenth century.

What is light?

Our interest in light began in a discussion one evening after Patrick had found himself in contemplative mood in his kitchen the previous day. He had been idly watching the washing-up water drain away when his eye fell on the sparkling colours shifting around in the disappearing soap bubbles. He decided to bring this observation to the group the following day. 'It must be to do with the light falling on the bubbles, surely?' ventured Sarah, the first commentator after he had described the scene. A torrent of questions ensued. 'OK, but why does it come out in different colours when the light is just coming from a yellowy-white light bulb?' 'Yes, and what determines the colours?' 'Why are they moving around?' 'Is it true for all soaps?' 'What if the light bulb had been red?' Gradually, as the questioning abated and attention began to focus on the basics, the simple question was raised: what is light? This ordinary phenomenon, an everyday feature of the natural world, suddenly seemed unfamiliar. The question silenced everyone. What, in fact, is it? Pushed to venture an answer, people were stumped: an unusual occurence in a normally chatty group.

Eventually a few first thoughts crept out. 'It seems to be linked to heat,' said Julie, always good at breaking the ice. 'Yes, that's true,' said others, as they began thinking about the Sun as such a dominant source of light – and fires and candles. 'And what about light bulbs too?' added Michelle. 'They're pretty hot, at least the old type used to be.' The conversation developed as the many different sources of light were imagined, most of which seemed to be linked to heat. 'On the other hand,' chipped in Sarah, 'some, like the newer LCD lights on bicycles and torches, seem cooler.' Thinking about the overwhelming dominance of the Sun, both directly through daylight and indirectly through plant life, prompted Julie to one of her characteristic left-field challenges: 'What would happen if you took the Sun away?'

Where does light come from?

Indeed the Sun really is the key to light on Earth. Even beyond its obvious role – beaming down the daylight and energising our solar panels – it's been working, for billions of years, building up the ancient plant life that gave rise to the oil and coal reserves that help to keep our lights on today. For us on Earth it's the Sun that counts. But let us not forget the rest of the universe: the aeons of space in which other stars burn as bright as our local star. So more strictly, light derives not just from the Sun, but from all the stars; and yes, it's our good fortune, here on Earth, to be living rather near one of them.

Light from this major source is indeed inextricably linked with heat. So perhaps the first and simplest thing we can assert about light is that it emanates from material sources, such as fires, candles, lamps and torches, and these are mostly hot (though in cases such as phosphorescence, light can be generated at lower temperatures). Most sources that give out light also give out heat and, in some cases, other related emanations, such as ultraviolet (UV) or infrared (IR). A second simple observation we can make is that generally light is emitted all around, in every direction. Some devices, such as spotlights and torches, are designed to be more directional, but the Sun, stars, candles and light bulbs are throwing out their beams in roughly equal measure in all directions. What reaches us in the direction of Earth is but a tiny fraction of what the Sun and stars are putting out.

'OK, so we've got the basics. It comes out from some definite material source and shines all around, in all directions. We've got that,' interjected Sonya, with a slight edge of impatience. 'But what actually makes it?' Sarah asked, taking a slightly more philosophical tack. 'Surely light had to be created somehow. How do we create light? Was the beginning of creation dark?' Ducking the tougher issue of the Big Bang for the moment, insight into how stars shine has developed enormously during the past hundred or so years. Direct observation of the spectrum of light from the Sun showed that it must consist of a vast mass of gas, mainly hydrogen and helium. In fact, it was in the light from the Sun that the new element helium was first detected in the 1860s – hence its name, from the ancient Greek Sun god Helios (Apollo).

So it was clear that the Sun is not a solid object like the Earth at all; it has no such familiar thing as a surface. It is instead a continually burning ball of gas, or more strictly plasma and gas. Plasma is the state that matter enters when it becomes so hot that the very atoms break up into their components. The negatively charged particles called electrons break free of the atoms to which they are normally attached, leaving the remaining atoms (now called 'ions') with a residual positive charge. At a temperature of some 6000 degrees this seething mass of particles, moving at extreme speeds, holds enormous quantities of energy. This gets released in the form of light particles known as photons (no relation to *pro*tons) as the various particles interact. It is this extraordinary outpouring of photons from the flaming periphery of the Sun that is what we know and love as sunlight. The strength of the light given off by each square metre of the Sun's surface is about 63 million watts; that is about a million household light bulbs for every square metre.

So now we get a sense of where the light that bathes our planet comes from. The flaming boundary of the gassy Sun sends out an endless and intense stream of photon particles into the void that is all around. But of course, as Sonya immediately appreciated, this energy radiating out from the outer layers of the Sun must itself have come from somewhere. What actually drives the Sun, what fuel is it burning?

The answer to this question in physics came not from astronomers probing the heavens, but from studies at the laboratory bench here on Earth. Marie Curie had been studying the curious properties of a naturally occurring mineral found in Germany, called pitchblende. In effect, by noting the way a piece of pitchblende lying around on a bench was able to blacken some distant photographic paper, she had discovered what we now know as radioactivity. Today we have a rich understanding of this extraordinary phenomenon. Radioactivity turned out to be a process of emission of a penetrating form of energy from the deep interior of atoms. At their very centre lies a tiny kernel known as the nucleus (Latin for 'kernel'). This kernel is perfectly stable for almost all the elements that surround us – oxygen, carbon, iron and so on – but, for a small number of heavy ones, it can break up, and when it does it releases vast amounts of energy. This is what Marie Curie was detecting in her pitchblende: a naturally occurring source of the radioactive element uranium. More significantly for the Sun and its energy, however, certain other kinds of nucleus are able to fuse together rather than break up, a process that also releases vast amounts of energy. This process, termed nuclear fusion, occurs for elements at the opposite end of the scale from uranium. Whereas the heavy elements tend to split, it's the lighter ones, hydrogen in particular, that are able to fuse.

The Sun, like other stars, is made largely of hydrogen, the lightest element. At its core, under the immense pressure of the mass of gas pressing inwards, the nuclei of hydrogen atoms are continuously fusing together, creating energy on an unimaginable scale. This is the sense in which the Sun is burning; its fuel is hydrogen, and it is nuclear fusion rather than normal combustion that generates the heat and light. This energy transfers gradually to the exterior regions of the Sun where it passes on to the photons that ultimately escape and reach us here on Earth. These photons *are* the visible light that surrounds us. But there are other photons too, ones with higher energy such as ultraviolet radiation and ones with lower energy such as infrared – neither of which we can see but which also beam down on us from the Sun. It is one of the special fortunes of Earth that the more harmful, high energy photons are mostly filtered out by the Earth's atmosphere.

This overview of the origin of light still left one of the group's questions unanswered: did light exist before the stars were formed? This was an interesting thought, revealing the extent to which basic astronomical knowledge seems to have gradually percolated into everyday thinking in recent times.

Now we know that there was indeed a time before the stars were born. We know that the universe was once filled with a thin haze of hydrogen gas with no solid lumps. Gradually stars condensed out of this gas, the hydrogen atoms being pulled toward one another by the gravitational attraction between them. This is how stars came into being, but in the period before this, the first billion years of the universe's existence, no large luminous objects existed at all. It is hard to imagine such aeons of time and to visualise a universe in this dark, diffuse condition. Nevertheless scientists and mathematicians today are studying the light that set out long ago from far distant galaxies and theorising about how the universe must once have been. It is because light travels at a finite speed, taking time to reach us, that we are able to look back in time in this way. The state of the universe and the form of its energy in its earliest moments is an active area of contemporary research.

Such esoteric concerns were not on the mind of Julie as this discussion about light took its next turn. 'Never mind the first few years of the universe and the light from galaxies astronomers can barely see. What about the light we see and use every day? You say light is associated with heat and comes from burning things like the Sun and stars, but what about the Moon? That's not on fire or hot, and yet on a cloudless night you can easily see by its light.' No sooner had she spoken than she realised she already knew the answer – the Moon merely reflects the light falling on it from the Sun. 'OK', piped up Michelle, keen to get back to something tangible, 'what does that actually mean, what is reflection?' Yes', added Mary, 'what about water? Water reflects light, so do other things like mirrors, glass, shiny things. Why do they?'

Reflection

As so often happens when scientific concepts get teased apart in discussion, there is hidden ambiguity in the way we use words, and 'reflection' is a good example. Ordinarily we associate the word with the images we see in the surface of a calm pond or a mirror. In ancient Greek mythology Narcissus, the son of a river god, saw an image of himself reflected in the surface of a pool. To his ultimate detriment he fell in love with it, not realising it was merely an image. Understanding what images are shouldn't be quite such a problem for us today, surrounded as we are by glass and mirrors. Clearly the connotation of the word 'reflection', as in an image in a pond, is distinct from its meaning as a process that deflects light from its path, for instance sunlight reflecting off the Moon. The common factor is captured in the original Latin word *reflexio*, which means 'bending back'. But what exactly is it that gets bent back?

In the case of the Moon, it is the light from the Sun hitting the Moon's surface that gets bent in all directions, including back towards Earth (Fig. 4.1). But what is it that gets 'bent back' when you see an image of yourself in a mirror? The obvious difference is that unlike a mirror the Moon's surface is, like most everyday objects, rough. It is not flat at all; bits of it tilt and point in all directions. Rays of light falling on such a surface are inevitably 'bent' all over the place, as in the diagram (Fig. 4.2).

'Rays' are a very useful graphical device for explaining how light travels, but it's worth remembering that they are really just a

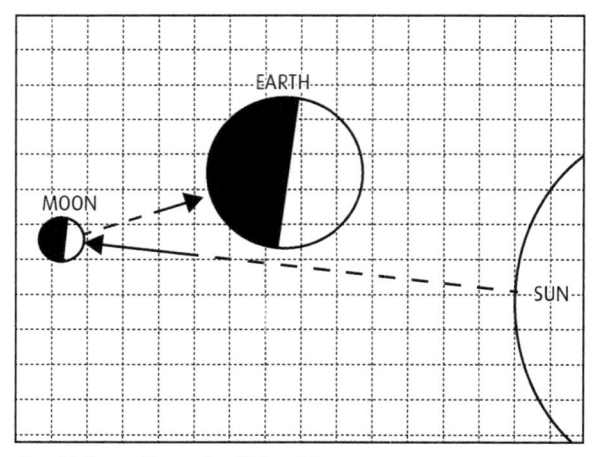

Fig. 4.1 Sunlight reflected off the Moon

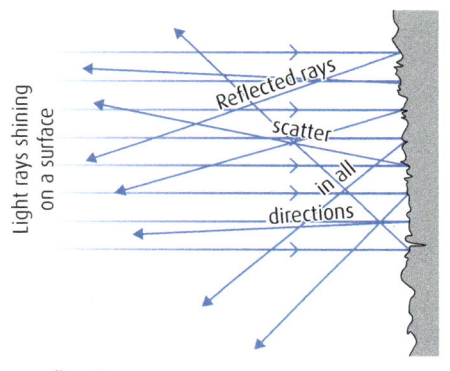

Fig. 4.2 Diffuse reflection

mathematical fiction – geometrical lines that give us a sense of how light behaves. In reality, light moves forward everywhere, not just along the lines we happen to draw. It is just this type of reflection, known as 'diffuse' reflection, that enables us to see objects at all, deflecting light from the daytime Sun and evening street lamps. Some surfaces, on the other hand, are particularly good at reflecting. The key feature of a mirror or the shiny exterior of a new car is, unsurprisingly, the smoothness of its surface. When a surface is seen to be smooth, not only to the eye but also at the microscopic level, rays of light will all be deflected in more or less the same direction (known technically as 'specular reflection'). Of course there is nothing magical about this; it's pretty obvious, since for a smooth surface, all parts of the surface are similarly aligned.

'Good, that's cleared up the basic issue about the Moon reflecting sunlight,' interjected Sonya, keen to avoid yet another interesting diversion threatening to take us even further from the original point. 'Can we talk about the main example of reflection – the image you see in the mirror every morning?' 'Yes', the others chimed in, in rapid succession. 'Why is the image in a mirror facing you?' asked one. 'It's reversed for some reason,' pointed out another. 'Do you realise the face you look at in the mirror every day is not the same one that everyone else sees? If you've got a wonky nose, veering to the right, you'll see it veering to the left.' This last point seemed a bit worrying: are you really the only one with the wrong impression of how your face looks? Sarah looked a little pensive as she tried to work out in her head whether we all get the same view when we stand in front of a mirror as a group. 'Does everyone see a reflection of me in the same place as I do?' she asked. Michelle took the point one step further, posing a question that harked back to previous discussions about evolution: 'Why haven't we evolved so our eyes are not deluded by reflection?' This seemed an interesting way of looking at the psychology of perception, and provided a good lead into the theory of images.

Images

Following our appreciation that light travels in straight lines, at least in empty space, and then gets bent when it reflects off a shiny surface, the way was now open to understand images. The trick developed by early scientists (or more strictly 'natural philosophers') was to construct fictional rays as geometrical lines on a diagram. These constructions help to answer many of the questions we have posed. Here you are, staring into a mirror at your face. Imagine the light falling on it from a light bulb or the natural daylight. This is being reflected off your skin and is beaming out in all directions. It radiates outwards from your face in every direction – that's how everyone else is able to see you. Naturally some of it travels in the direction of the mirror and hits it at some angle. These particular rays of light will, of course, be reflected off the mirror. What is more, as you know if you have ever shone a torch beam at a mirror, the light bounces off at exactly the same angle that it hits at.

Now imagine a group of diverging rays, all starting out from the tip of your right forefinger (A in Fig. 4.3 below) and being reflected by the mirror. If your eye was in the position shown, looking at the mirror, it would pick up the diverging rays of light, but would be fooled. The brain, linked to the eye, religiously follows the rule that light just travels in a straight line from its source. So the deluded brain interprets the light entering the eye as though it came from behind the mirror (the position marked 'A').

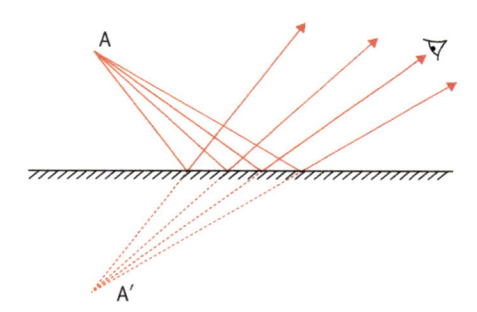

Fig. 4.3 An image seems to be behind a mirror

So I am sorry to say that, despite its enormous sophistication, the human brain doesn't do bent light rays. It prefers instead to construct fictitious images when looking into mirrors – the point raised above by Michelle, wondering why evolution hadn't sorted out this delusion aeons ago! To explain why your face in the mirror does not appear to be exactly the same as the face other people see, we need one more diagram – a slightly more complicated one.

The question is: why does your image in a mirror seem to have a left hand where you have a right hand? Why does the ring in your right ear seem to be in the left ear of your mirror image?

In this diagram (looking down from above), light from the different parts of the arrow (1) is reflected off the mirror (3) and reaches the eye (4). But the brain assumes the light came in straight lines, so forms the image (5) in which the tip and tail of the arrow are reversed. It's the same when you look in the mirror. The image of you is turned through 180 degrees. If you were facing north your image is facing south, yet your right hand is still on the right of the image. So the imaginary person facing you in a mirror appears to have a left hand where the real person has a right.

Thinking back to the ill-fated Narcissus, Michelle realised that images in which left and right appear to be reversed is not only true for glass mirrors. You see it when you lean over and look into a pond; you see it in countless paintings where reflections are painted in to create the impression of a watery surface. 'Why does water reflect?' she asked spontaneously, then quickly realised she could answer her own query. All it takes is a smooth reflective surface to ensure an image appears,

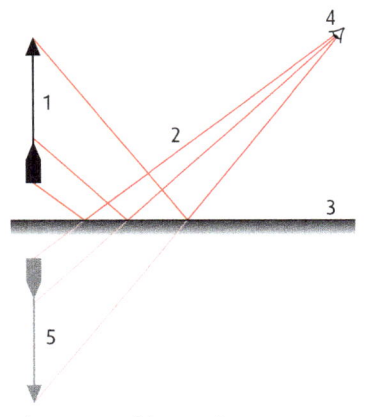

Fig. 4.4 The image is reversed in a mirror

and water, left to settle, will arrive at a smooth, flat surface if undisturbed. So it becomes a mirror as far as light is concerned, bouncing back the rays that fall on it, some of which end up in the eye of the beholder. A surface of water, or indeed any smooth surface with reflective properties, acts just like a mirror in this respect. The interesting case of surfaces that are shiny but, despite this, don't reflect an image – such as polished marble for instance – take us into the deeper nature of materials, highly interesting but...

'Can we leave polished marble for a moment and get back to the subject?' A heartfelt plea from Julie. 'These ray diagrams and inversions of left and right and fictitious rays and imaginary images are doing my head in. We only got on to the topic of reflection because someone mentioned the light from the Moon and how it wasn't from anything burning on the Moon, just reflection of the Sun's light.' 'Yes', jumped in Sarah quickly, 'can we go back to talking about light from the stars? If it comes from a star far, far away it must have further to travel, so surely it will lose energy en route?' 'What about ultraviolet radiation too?' added Michelle. 'I had thought it was a short wave, so how can it reach us from the Sun so far away?' – recalling, somewhat shakily, a previous discussion about waves. The concept of light as a kind of wave contrasts with the earlier description of it as a stream of particles (photons). The disparity between these two models of light is taken up later in the chapter.

Light from afar

This area of discussion brings to life a very powerful, fundamental concept. It's about how things fade out over long distances. Its relevance to everyday experiences could hardly be stronger. Listening to musicians, the farther away they are the softer they sound. Using your mobile phone in the countryside, the signal may have faded to nothing. Or, taking the opposite case, when you drive at night on a two-lane road, oncoming headlights that are tolerable far away can be overwhelming as they approach. In all these situations we know that beams are stronger when you are close to the source and weaker as you recede from it. As these random examples show, it doesn't depend much on the nature of the beam. In one case it might be sound, in another, light; it could be the television signal from a transmitter or X-rays from a hospital generator. Either way, the intensity weakens as you move away.

Intensity is a useful concept in discussing this effect. After all, much depends on how intense the original source is, whether this is the amplifiers of a rock band, the flaming surface of the Sun or the transmitter on a mobile phone mast. The amount of energy being pumped out every second determines how powerful the source itself is. Clearly a lighthouse outranks your bedside lamp, and the Sun outranks both. But what about the stars: are they necessarily weaker than the Sun just because they look much dimmer?

The diagram below explains how the intensity of radiation from any source weakens as it travels outwards. In passing, having introduced the word 'radiation' into discussion, it is worth pointing out that the word has a neutral and very useful connotation in science compared to the terrible association it has acquired in everyday parlance. It is simply a word used to describe the way energy moves outwards from source, radiating in all directions – along a radius. So there is light radiation, heat radiation and sound radiation, not just the nuclear kind.

In Fig. 4.5 below, the red lines are symbolic, indicating a small sample of rays radiating from the source (marked 'S'). At the source the nine rays represent the particular strength or intensity of the source. Were there to have been 18, it would have been twice as strong.

By placing imaginary grids at equal distances along the beam, we can see how the beam gets steadily weaker, simply as a geometrical necessity. In the square closest to the source all nine rays pass through. In the second only three pass through each little square, and in the third only one. The beam is weakening from nine units in each square to three

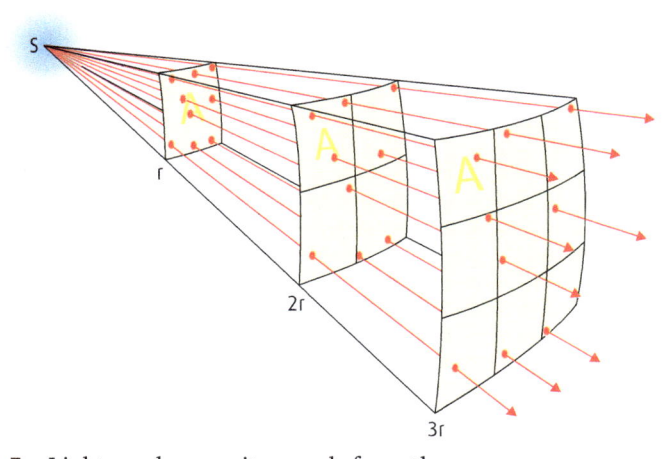

Fig. 4.5 Light weakens as it spreads from the source

THE DUAL NATURE OF LIGHT

to one, over equal distances. This shows that the weakening of a divergent beam such as this is a purely mathematical effect, not caused by any actual energy loss on the way. It's inevitable, even if there are no substances in the way filtering the beam. The light has to spread over an ever greater area. So the brilliance of a source of light, whether emitted from a lamp or a star, will steadily dim as you get farther away from the source.

The diagram actually tells us something even more powerful – it shows that the weakening effect is not just linear, but goes as the square of the distance. In other words, it is rapid; not just from nine to eight to seven, but from nine to three to one. This is known technically as the 'inverse square law'. As a caveat, we ought to acknowledge that not all beams of light progress equally in all directions. After all, torches and lighthouses and spotlights, for example, are specially designed to provide a beam that does not diverge too much; they use special mirrors and lenses for just this purpose. There's always the possibility too that a beam can diminish as a result of some substance in its path, its strength gradually diminishing not just by the geometrical factor, but also by absorption of some of its energy. Think of the Sun on a misty morning, or the sound of the television in an adjacent room, muffled by an intervening wall.

Before we allow this fascinating topic and the stream of questions it raises to divert us too far, let's get back to our original questions: the Sun and stars. We have now established that Sarah's point, that light from distant stars becomes less and less intense as it travels towards us here on Earth, is indeed true. Now we know why: it simply has to spread itself over a larger and larger area, and as a result each part gets less and less of the original energy from the star. That's why the stars are dim: their energy has to be spread so widely on their extremely long journeys. The Sun is brighter because its light has had so much less distance to cover. Some stars are actually far more luminous than the Sun, others are less so. The brightness we see is a result of both their original luminosity and the distance their light has had to travel to reach us.

The issue of ultraviolet (UV) light, brought up earlier by Michelle, has a simpler answer. The fact that this kind of radiation is known as 'short wave' is a bit misleading. It's not really the wave itself that's short – it still reaches comfortably all the way from the stars to Earth. Indeed, waves in their entirety can be of any length. What is short about ultraviolet waves is what is defined as the wave*length*. This is simply the distance from one peak to the next within a wave, which can vary enormously from one type of wave to another. Radio wavelengths are relatively long and X-rays are short, while visible light is somewhere in between. In the case of UV, the wavelength is referred to as 'short' because it is shorter than visible light that accompanies it in sunlight.

Below is a diagram showing how the colours of light fit into the wider spectrum of waves (Fig. 4.6). These various types of wave, from radio to gamma ray, differ only in their wavelength. Collectively they are known as 'electromagnetic' waves, to distinguish them from other kinds of wave, such as sound or water waves.

So UV, just like visible light, travels across the vast universe, spreading its energy ever more thinly as it does so. Fortunately much of it gets filtered out by interaction with molecules when it hits the Earth's atmosphere – just one of the many extraordinary circumstances that make the Earth conducive to life in a potentially hostile environment.

This excursion into fundamental physics seemed to have gone far enough for Sonya at this point. Something a little more personal was taxing her. 'It was my 40th birthday last week,' she announced with some pride. 'Could I find a star whose light that we see today actually set out on its journey to Earth when I was born, 40 years ago?' An interesting question, if a little subjectively posed. It stimulated thoughts about just how far away the stars are and how many of them there are. Can it really take 40 years for light to travel across space to reach us? Are there really so many stars that, statistically, there is bound to be one exactly 40 light years away?

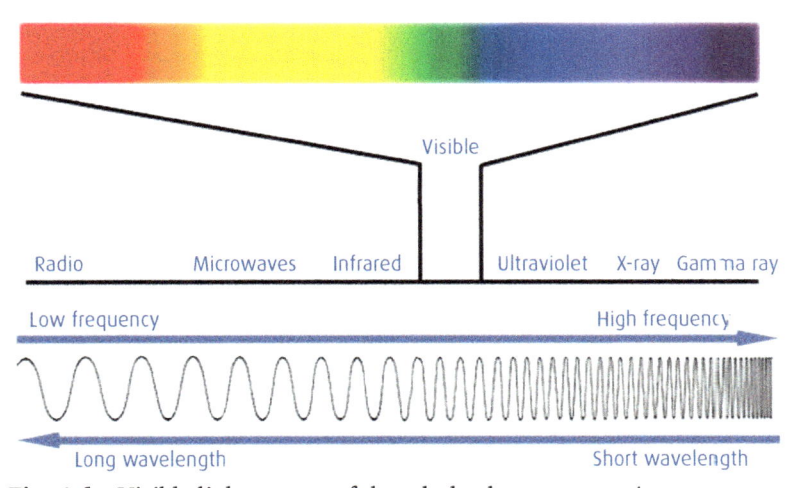

Fig. 4.6 Visible light as part of the whole electromagnetic spectrum

Stars are indeed extremely far apart compared to the humble distances we are accustomed to in the micro-world of human existence. The question of how fast light travels has divided thinkers since ancient times. Kepler, for example, thought it must travel infinitely fast since empty space presents no obstacle to it, whereas Galileo proposed an experiment, carried out in 1638, to measure the time delay when a veiled lantern one mile away was uncovered. The delay, which we now know would have been 11 microseconds, was, not surprisingly, undetectable at the time. But in 1676 the first reasonable estimate of the speed of light was made by the Danish astronomer Ole Rømer. He cunningly compared measurements of the arrival time of light from one of Jupiter's moons taken when Jupiter was close to the Earth with those taken when it was far from Earth. He realised that the light took longer to reach us when Jupiter and its moons were farther away, and was able to estimate its speed.

Today we know that light does indeed have a finite speed. What is more, we know from Einstein's work that this represents a limiting speed that no material object can ever achieve. The speed is approximately 186,000 miles each second or, to be more precise (and to convert to metric), 299,792.458 km per second. So light travelling at this speed for 40 years would have travelled 9,454,254,955,488 km; let's say around 10 million, million kilometres. According to an Atlas of the Universe, there are three visible stars that are around 40 light years away; for each of these there are at least nine others invisible to the naked eye. So, yes, Sonya, it's just possible, but not certain, that there is a star whose light shining on you today actually set out on the day, or at least in the year, that you were born!

This revelation about starlight travelling for years, even millions of years through the vastness of space sparked off an even more profound realisation. 'Are we saying the light has travelled through emptiness all the way since it was created? Was there no air or anything else in the way?' many asked. Yes, it may seem amazing, but it's true that no material thing at all exists between a star 10 million, million kilometres away and us here on Earth (apart from the tiny 100 kilometre layer of the Earth's atmosphere). 'Are you honestly saying that a photon of light emitted from that star has kept on going for 40 years at the speed of light, uninterrupted till it hit my eye? Am I really the first thing it's ever encountered?' Yes, in principle that's right, space is essentially empty: no solids, no liquids, not even gas. To be strictly accurate, we ought to acknowledge that there is a very, very thin presence of hydrogen atoms in outer space, just a few per cubic metre, compared to ten trillion, trillion in Earth's atmosphere.

Models of light

So how does light actually make its unceasing journey through space? Michelle, looking somewhat baffled as she contemplated this question, got nods of approval as she queried: 'Surely it would run out of energy somewhere along the way?' This question of how light actually progresses through space has proved baffling throughout the centuries, and not just to lay people. Natural philosophers argued for hundreds of years over alternative explanations. Today we are forced to accept there are in fact two quite distinct ways of looking at this, two versions of a model for light.

During the eighteenth and nineteenth centuries it was more or less proven that light travelled in the form of a wave – a kind of oscillating disturbance, analogous to the ripples on the surface of a pond. Many experiments over centuries had provided evidence for this model. Furthermore, in the mid-nineteenth century it had been established that these oscillations were not in fact ripples of a material kind, as waves are in water, but rather of tiny electrical and magnetic influences (dubbed 'fields') that rise and fall in strength millions of times each second. These waves (electromagnetic, as they were named) were able to travel without any medium to support them. They could travel in glass or water, or air, but also in a vacuum. With this model in mind, we can say that light from the Sun or a distant star travels as the vibrations of a wave that was set up in the star's outer regions; it then travels outwards, spreading over an ever greater area.

However, with the arrival of the quantum theory at the beginning of the twentieth century, different evidence was thrown up by experiments suggesting that light does not travel as a wave but in the form of a stream of particles (called photons), each carrying a tiny quantity of energy but having no mass at all. With this model in mind, the stream of photons originates in the star and continues travelling without loss of energy forever, unless it should chance to hit upon a stray planet or dust particle – a most unlikely occurrence. The uncomfortable co-existence of two competing models for light is a reality that scientists were forced to come to terms with during the twentieth century. The so-called 'wave-particle duality' is a fundamental aspect of the quantum theory, and implies that each model for light explains some aspect of its behaviour but not others. A single, all-embracing model is not available to us. This issue of ambiguity is taken up in chapter 5, where we discuss the nature of models themselves. Whichever model is adopted, it's clear that light does not 'run out of energy' as it travels, but simply spreads itself ever more thinly over ever larger regions of space.

The colour of light

At this point the path of discussion took an abrupt turn. Ignoring the profound question of the nature of light – photons or waves – the very idea of light hurtling through the endless void of space inspired a simple but penetrating question from a puzzled looking Julie: 'Can you actually *see* a light beam in space?' she asked. 'Let's say you switched on a torch beam, what would you see? Would it throw out a beam as you'd see here on Earth with a torch or searchlight?'

Trying to imagine this unlikely situation provoked a cascade of thoughts. What does outer space actually look like? Would there be sky? How do we see a torch beam here on Earth? A picture gradually began to emerge as people pieced together various bits and pieces of knowledge that came to mind. There's no air in space, no atmosphere, no sky, no clouds – no material substance at all. When we see a torch beam, in reality we are seeing it light up some dust particles or droplets in the air, like a searchlight shining up into the sky on a misty night. What we see is light reflected back to our eyes from these tiny things. Out in space, however, there aren't any dust particles – there's simply nothing, an emptiness we can hardly imagine. So, strange as it may seem, we would *not* actually see the beam of light from a torch out in the void. Paradoxically, light itself is essentially invisible. In fact you'd only see a beam of light were it to be shining directly into your eyes. If we looked at it from the side, it would be invisible! What is more with no air, no substances at all, there would not even be a sky; there'd be nothing to interact with the sunlight to produce the blue colour. It's a strange truth, but space is black. Even the Sun itself loses its iconic colour out in space, bleached to its original white, a pallid blend of all the colours of the spectrum.

In characteristic fashion, the onward march of explanation came to a juddering halt at this point in the discussion. 'Hold on a minute,' a familiar cry from Julie, ever vigilant for scientific sleight of hand, 'what justification have you got for claiming space is black? Why is the sky blue then?' Inspired by these radical challenges, Sarah joined in. 'Is there any reason the sky must be blue; could it be green for instance?' 'Anyway', added Michelle, 'it's not always blue, is it? Haven't you ever seen it pink or grey, for example?'

These exchanges heralded a new moment for thought, this time about the colour of light. Despite believing that hardly anything from their school science lessons had stuck, everyone in the group seemed to know that light from the Sun, though apparently whitish-yellow, was in fact made up from a whole spectrum of brilliant colours, the colours of the rainbow. With a bit of reflection, it was realised that it must be the air around us that affects the colour of daylight. Michelle recalled that sunsets were often more of a brilliant red and flaming orange in the polluted atmosphere of industrial regions. Sarah pointed out that these colours tend to dominate at the beginning and end of the day. From such everyday observations they saw there must be some link with the state of the atmosphere and time of day. 'Come on, Andrew,' they chivvied, 'what's the explanation?'

These observations from everyday experience go a good way towards explaining the story. Light from the Sun is indeed a blend of all the colours of the spectrum. What the eye and brain detect as different colours are simply light waves of slightly different wavelength. The redder colours come from waves of longer wavelength, the yellow and green, middling wavelengths and the blue and violet, shorter ones (nearly half as long as red). As you see in a prism, these colours can easily get separated when they pass through some kind of transparent medium. Light of different wavelengths gets bent to different degrees as it passes through something such as glass. The wavelength that gets bent the most is blue, while red is bent the least. A related process occurs in the atmosphere, another transparent medium, though in this case a gassy one rather than liquid or solid; it's not just empty space, however much it may look it to us. In fact it's filled with molecules of nitrogen, oxygen and carbon dioxide. Light from the Sun interacts with these molecules in such a way that a portion of the light gets re-radiated in all directions, not just the direction in which it was originally travelling. The colour that gets scattered in this way the most is blue. As a consequence of this scattering, light waves at the bluer end of the spectrum appear to come at our eyes from all directions; the other colours, less so. This creates the illusion of blue light arriving from the entire hemisphere above us – what we call the sky.

With these new insights in mind, discussion began to uncover some of the other colour effects. As we look at the Sun, for example, the bluer part of its light has been scattered away more than the other colours, so it leaves behind the orangey-yellow mix of colours that we associate with it in daytime. In the evening and early morning, the Sun is so low in the sky that its light has skimmed through many more miles of atmosphere on its journey to our eyes than it has at midday (Fig. 4.7). Almost all the bluer colours and some green have been scattered away entirely, leaving an ever stronger element of red in the mix. Hence we see a gradual transition through deeper yellows and oranges to red in the first and final minutes of the day.

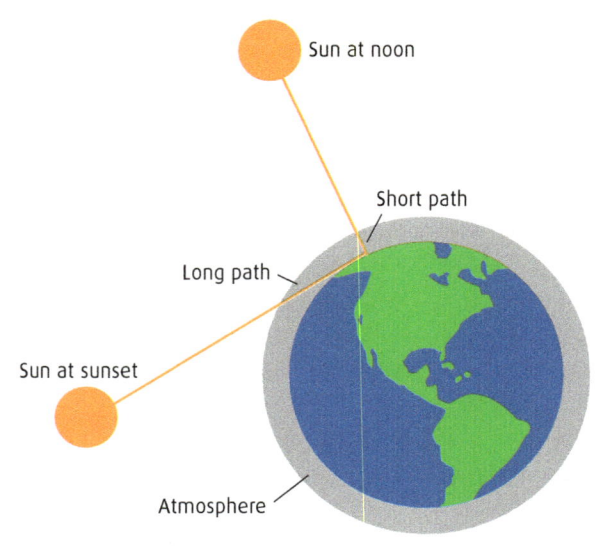

Fig. 4.7 Sunlight passing through the atmosphere at noon and evening

The atmosphere, as well as supplying the essential gases of life and protecting us from harmful radiation, is also responsible for some of our most profound aesthetic experiences: scudding clouds, flaming sunsets and the ever-changing hues of the sky.

Conclusion

As so often happens in open-ended discussions, this excursion into the nature of light has taken us through many fascinating aspects of the world around us and illustrated many profound and useful concepts. At the same time it's probably left more questions hanging in the air than it's answered. Perhaps the most perplexing of these is also the most basic: what exactly is the nature of light? This is the question originally posed following a simple comment about colours in soap bubbles at the kitchen sink. A clear and straightforward request – what is light – met with a singularly evasive answer: 'Well, it depends'. Sometimes the behaviour of light can be explained as a wave, with different colours corresponding to different wavelengths, and these waves being bent (or refracted) to differing degrees when they pass through transparent substances like air and water. But sometimes the behaviour of light is best explained as a stream of tiny immaterial particles, shooting out from the Sun and light bulbs, getting absorbed by particles in the air or scattering off them.

You'll be glad to know this apparent duality in the nature of light not only baffles us; it has also troubled scientists and philosophers throughout modern times. In the early twentieth century it betokened the break-up of classical ideas about the nature of the physical universe and became the wellspring of the radical new quantum theory, whose discomforting implications remain with us today. The implication of this kind of ambiguity forms the theme of the next chapter.

5
Models

'Is he actually *seeing* the particles?' demanded Julie, reflecting on a spell-binding discussion the group had just had with a particle physicist working with the Large Hadron Collider at CERN in Geneva. Being able to discuss a question like this face to face with a researcher is one of the great benefits of visiting a lab in person. Under questioning from Mary the researcher had revealed that what he was 'seeing' was in fact a long stream of digits flashing past on a computer screen. By analysing these, he was able to determine the presence or otherwise of a fundamental particle.

This exchange about what is 'real' is typical of a recurring theme in discussions about science. Scientists routinely talk about things that are models of reality as though they are the real thing; to them a computer representation of the effects of a particle in a collider *is* the particle. Scientists get used to this kind of metaphorical talk, brushing aside any unease about the distinction between theoretical construct and 'objective reality'. For most people, however, science is generally understood in more concrete terms. Take the force due to gravity, for example, or the structure of DNA or the flow of electric current: these are taken to be hard-edged realities in the natural world, revealed to science through experimental discovery. The idea that they are more accurately understood as *models* of the natural world – developed over time by human beings, based on meticulous observation and experiment – comes as something of a surprise. The role of models in explaining observations is a regular discussion point in science groups, often starting from a question about what it actually means to 'see' something.

Facts, formulae and laws are generally perceived as key ingredients of science. The reality described by science is commonly understood as being definitive rather than open to interpretation. The more tentative concept of 'modelling' reality appears less often in scientific discussions than it does in economic, political or social ones. We talk of a progressive model of taxation or a zero tolerance model of policing. My hunch is that this difference suggests an unconscious conviction that natural sciences describe an objective reality, whereas social sciences are deemed more subjective and conjectural.

Outside the intellectual sphere, the word 'model' has a wide range of uses. The earliest dates from the sixteenth century, when the word denoted a small-scale version of something. A little later it came to be associated also with a thing or person to be imitated – the sense in which it is still used today in the phrase 'role model'. With the advent of models in fashion and car production, the word has come to be associated with the ideal, a sense that extends to contemporary idioms such as 'model student' and 'model parent'. The *Oxford English Dictionary* definition that seems most appropriate for science today is of a 'simplified description, especially a mathematical one, of a system or process, to assist calculations and predictions'.

The way of science

Scientific endeavour, as we understand it today, starts by finding ways to measure things. I remember a science teacher at school telling me this at the tender age of 14 and my thinking: how boring does this get – is it really just about tape measures and thermometers? At the time (the 1960s) I was keen to explore the mysteries of the Sun and stars, and to learn about exciting new things such as heart transplant operations and electronic computers. It took many decades for me gradually to appreciate that each of these topics had in fact come to be understood through careful measurements – of light intensity, blood flow and voltage, for example.

Good measurement leads to good data, and good data is the basis for useful models. However, measurements are never without some degree of inaccuracy and imprecision; developing models from data necessarily involves a level of interpretation and uncertainty. This issue of imprecision and uncertainty came to the fore in a discussion one day about disease during the 'bird flu' scare. Discussion moved on to the relationship between health and poverty, and it was decided to research the topic online. We found the following graph, which shows the relationship between infant mortality rates and inequality of income (Fig. 5.1). It neatly illustrates the difference between a mathematical model and real data.

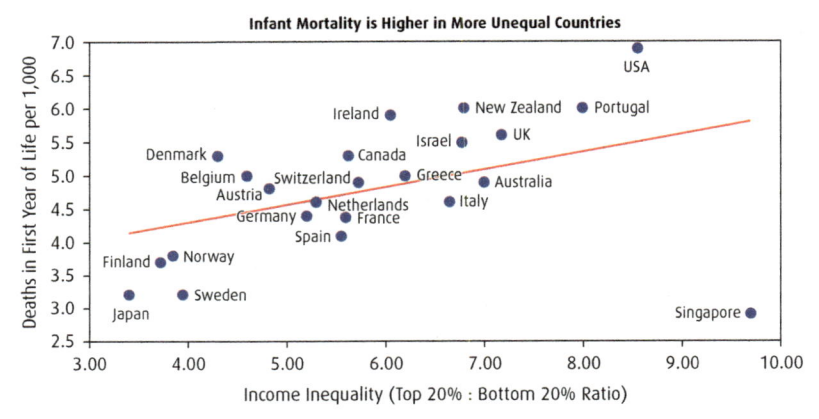

Fig. 5.1   Modelling a relationship

The scattering of data points, though not conforming perfectly to a precise pattern, reveals a clear tendency from which a relationship can be interpreted. Statisticians employ a mathematical procedure to find the line that fits the scattered points best. This graph provides an example of how measurements can suggest a model – in this case, one in which the infant mortality rate is linked to levels of income inequality – even when the data don't suggest an exact rule.

Perfect and imperfect models

In many branches of science idealised models of phenomena are developed using a simple and precise mathematical formula, even though the real situation is more complex. This is one way in which the connotation of a model as 'perfect' or 'ideal' applies in science. An interesting and explicit example of an idealised model occurs in understanding the behaviour of gases. Measurements of the pressure and volume of gases at various temperatures, made in the eighteenth century, led to two perfect mathematical relationships known as Boyle's Law and Charles' Law. Later, when it was possible to take measurements at much higher pressures and lower temperatures, these laws no longer held so perfectly. As result 'real' gases were said to behave like 'ideal' gases only under normal conditions. In discussing this point, Sarah spoke for many when she asked, perfectly reasonably: 'What's the point of an idealised model if it doesn't accurately represent what is going on?'

The reason is rather utilitarian: models prove valuable if they enable predictions to be made that prove to be reasonably correct sufficiently often. They may idealise some aspects of a complex reality, but if they explain experimental results well enough they are adopted (until a better one comes along). Thus, for example, a model of the spread of an epidemic or progress of a hurricane potentially enables the future path to be predicted. If it does so better than pure guesswork, it is probably useful and may save lives. As Julie, with her background in sociology, put it in a discussion one day: 'Scientific truth is only as good as its usefulness – if it works, it's science.'

A large part of what scientists actually do involves developing and using models: finding mathematically exact curves that closely match reality, then using them to calculate things such as the rate of digestion of food or the rate of decay of nuclear waste. By this means theoretical models enable any number of situations to be dealt with, rather than each one requiring separate consideration and measurement.

As anyone who relies on weather forecasts knows, models cannot guarantee accurate results in all circumstances. Their efficacy depends on whether the benefits gained from their accurate predictions outweigh the drawbacks of their inaccurate ones. Modern weather forecasting is based on highly complex modelling using extremely powerful computers. It offers clear advantages overall to farmers, sailors and concert promoters, among others, even if it occasionally fails them. Sometime it is not just inherent uncertainties, as in weather forecasting, that blight a model; flaws in the underlying theory may result in failure at a more profound level. In a discussion about the history of genetics one day, Julie pointed to the eugenic model, which had predicted social catastrophe because of the differing birth rates of different social classes. But, she recalled, it failed to take into account the effects of female emancipation on birth rates. Trying to determine which factors to build into a model is far from simple, as much in science as in economics or sociology. Models are of central importance to both the social and the natural sciences though, as Julie observed wryly, 'perhaps they last a bit longer in the natural sciences?'

Limits of modelling

Models by themselves cannot tell the whole story of a phenomenon. They produce a result when given some data, but may or may not offer an explanation of the process in between. At the simplest level they may simply predict the chances of getting a particular result given some starting point, rather like predicting the odds on a racehorse based on its previous form while knowing nothing about its health or capacities (my late aunt always reckoned a close look at their knees would predict the Grand National winner). An interesting example in science comes from the history of genetics. The monk Gregor Mendel is credited with founding modern genetics, as a result of his cross-breeding experiments with peas in the gardens of the Augustinian Monastery of St Thomas in Brno in the 1860s. His work showed that traits were inherited according to statistical rules, and he invoked the idea of 'invisible factors' to explain these rules. The actual mechanism, involving genes located in the DNA molecule, was not established for a further 80 years. There is often a time lag between an initial model being useful as a predictor and the emergence of a later version capable of describing a mechanism. Aspirin is another example: despite being developed as an effective drug in the 1890s, its actual mode of operation was not confirmed until the 1960s.

These examples demonstrate a key feature of scientific activity: the way in which models build over time, often involving several contributors. New ideas, techniques and resources enable increasingly sophisticated explanations of the mechanism of action to develop over time. In the case of weather forecasting, for example, the growth in methods for collecting data across the globe and in computing power for processing it has led to increasingly sophisticated modelling and forecasting. However, as everyone knows from experience, modelling often involves a degree of uncertainty: storms can, and do, alter course unexpectedly.

In rare cases, new data or radically new ideas may simply overthrow an existing model. Examples have occurred famously throughout history. In the early sixteenth century Copernicus proposed controversially that the Sun, rather than Earth, was at the centre of our system of planets. In the late seventeenth century Newton made a revolutionary leap of imagination in showing that the gravitational force that causes objects such as apples to fall to the ground also keeps the Moon in orbit round the Earth. This kind of dramatic remodelling can be very disruptive, causing serious disputes in the scientific community and sowing confusion outside it. From our modern perspective, however, these revolutionary moments serve to remind us of the provisional nature of models. We need them, we use them, often we rely on them – yet we know from experience that they may need to change as new information and new ideas emerge.

Reflecting on those extremely rare occasions in which we fundamentally reframe our understanding of the universe causes us to think carefully about the limits on the role of science. Science is the art of measuring and of developing models to produce predictions. Discussions about scientific models often provoke deeper questions about why the universe is in the condition it is in. What actually is electrical charge? Why is there a gravitational pull between objects? Why do we have the particles we have and not others? Science presses hard on these questions, closing in as far as it may, but ultimately questions about *why* things are as they are, as opposed to *what* they are, lead us into the realms of philosophy and religion. Science can clear away much of the falsehood on the way, but the ultimate questions of reality call more for acts of belief than scientific explanation. It's a matter of opinion how far scientific models alone provide a full and satisfactory explanation.

The development of models

Models do not generally simply appear, quietly making themselves known to the occasional genius. Historically they have often emerged in a dialectical manner: an initial idea develops from observations, later giving way to a more convincing one as new data or ideas emerge. The ancient Greek scientist Empedocles contributed the idea that all natural substances were compounded of a few elemental substances: earth, air, fire and water. This was an important advance over the previous notion that each substance is composed of its own unique element. The model was gradually developed by experimentation, particularly after the Renaissance. It then retained the ancient Greek concept of elements, but abandoned the particular ones proposed by Empedocles.

A further example is the development of the concept of combustion. A fire-like element, inherent in flammable substances, was conceived of by seventeenth-century natural philosophers to explain the process of burning. This supposed 'phlogiston' was believed to be released during combustion. Well-intended though this explanation was, the model simply didn't stand up to experimental tests. When oxygen was discovered in the 1770s a completely different model of combustion was developed, one that we still use today.

Models can also develop in less disputatious ways – not so much through fiercely fought rivalries, but more through gradual accumulation. A recent example is the development in science's understanding of the host of fundamental particles of which all matter is composed.

These have been investigated continuously since the early twentieth century by forcing tiny particles to collide with one another at immense speeds in powerful machines. With each decade, new and often baffling discoveries were made by physicists around the world. Each new result didn't exactly contradict previous ones; it simply added to an increasingly fragmentary and confusing picture – a 'menagerie of particles', as it is sometimes called. By observing patterns in the information as it accumulated it has proved possible for scientists around the world to agree on a so-called 'standard model', based on the results of countless experiments. Nobody pretends it is the final word on the ultimate constituents of matter, but it provides an enormously helpful tool to rule out some hypotheses and lend support to others. The 'standard model' is expected to continue developing as new technologies, such as the Large Hadron Collider, probe ever more deeply into the behaviour of fundamental particles.

Imagination often plays a crucial role in the development of theory. Einstein famously used his imagination to think about the speed of light and the nature of gravitation. In one of his thought experiments he imagined riding along at the speed of a light next to a light wave. He realised the wave would have to appear stationary to him as he travelled alongside it. This seeming impossibility caused him to question the entire model of space and time, resulting in his special theory of relativity.

Multiple models

How intellectually reassuring it must have been in times when the mechanism of the world could be explained with certainty, when belief was absolute. There was earth, air, fire and water...period. The Earth was the centre of God's universe and Earth-like things fell towards it. The heavens revolved on rigid, transparent spheres and heaven-like things rose up towards them. Offering such confidence and clarity, it is no wonder that Aristotle's explanations endured for centuries, millennia even.

However, the radical new concepts of observation and experimentation that developed during the Renaissance would ultimately spell an end to this vision. The dramatic increase in laboratory testing during the eighteenth and nineteenth centuries saw a steady growth in understanding across the natural sciences. The motivating idea was to reveal the precise and intricate workings of a created universe. The advance of theory appeared to be gradually drawing back the obscuring veil on an objective reality. Developing man-made models was not the general view of the scientific process; discovering God's truth was perhaps closer to prevailing beliefs.

But just as empirical advances grew steadily, so too did doubts. In the early eighteenth century Newton developed an evidence-based theory to explain the nature of light, published in 1704 in his book *Opticks*. He saw it as a kind of stream of particles ('corpuscles' he called them) which would be reflected off surfaces and refracted through transparent materials. A century later his fellow countryman Thomas Young also experimented with light, demonstrating that it cannot be particle-like but must be some form of wave motion. His theory was no idle speculation – it was based on evidence from ingenious experiments with narrow slits and lenses which showed light behaving rather like water waves, but with very much shorter wavelengths (over a million times shorter). The rival theories were forced to compete, as were theories about combustion, heat transfer, blood flow and a host of other contested topics.

Alternative theories multiplied and the expectation was that a winner would eventually emerge out of the competing models to vanquish all others. The apotheosis of this belief seemed to be James Clerk Maxwell's all-encompassing theory of electricity and magnetism in the 1860s. But as this chapter has shown, the earlier conviction that perfect theories would reveal an ultimate truth has slowly given way to the idea of models acting as more provisional concepts, aids for explaining what is observed and for making testable predictions. We have had to come to terms with some limitations: models are generally imperfect at any given moment, they change and adapt as new information emerges, and they usually have a limited shelf life.

Expectations were forced to change around the turn of the twentieth century when, in a somewhat esoteric area of research, the shining of ordinary light on to metals began to produce some anomalous results. This so-called photoelectric effect couldn't be reconciled with prevailing theory. It was Einstein's explanation of this effect in 1905 that lead to an even further fall from grace for the idea of the perfect model. Not only were models often found to be imperfect and provisional: sometimes they also proved to be Janus-faced.

Einstein had effectively unleashed a quantum model of light – a modern version of Newton's corpuscular model – just at the time that Maxwell's wave model had become firmly established. The unease introduced by this conflict between light as quanta (discrete particles) and light as a wave drove the leading physicists of the period to seek reconciliation between the two competing models. Each model appeared to explain some of the observations very precisely, but was unable to explain others. As these attempts at reconciliation failed, one by one, to provide a unitary explanation, a new era of ambiguity opened in science: the two models simply had to coexist. It was us, the conceiving minds, that would have to change, rather than the models themselves. We have had to accept that in some instances one model may have to share its explanatory power with other rivals. Light behaves like a wave when we look at it one way and like a particle when we look in other ways. We are forced to change our ideas about what a model is. In some situations the idea of a single version of a model of reality, compatible with all observations, may simply not be meaningful. Is it possible that the ways of nature may not be entirely susceptible to the way our minds conceive?

A discussion was once inspired by a comment from Sally, who had previously attended an adult education astronomy class. She had picked up the idea that perhaps there are concepts for which the human mind may simply not be ready. This provoked lively discussion about the nature of the human mind. 'I suppose the brain has evolved biologically like all the other organs,' said Sarah, observing that 'it seems to rely heavily on visual images'. It is true that much of our scientific understanding does develop through use of metaphors and visual imagery: an electric 'current flows', DNA 'encodes', amino acids are 'building blocks'. But in trying to understand contemporary physics some things are simply not conceivable in the ordinary sense – satisfactory pictorial images may not exist. For the physicists, who nevertheless need to model the fundamental processes of nature, pure mathematics becomes the key to explaining the strictly inconceivable.

Coping with ambiguity

Interestingly this more liberated concept of models in place of rigid laws came as something of a relief to members of one discussion group. Julie recalled how her teachers would sometimes commence a new school year by dismissing whatever explanation had been given the previous year as being too simplistic or even misleading: 'No, there aren't really shells of electrons inside atoms, as you were taught last year. Instead there are clouds of negative charge.' Mary described how her daughter came home one evening complaining that a science teacher had told her to forget what she had been told last year, because they would now learn a different explanation. 'Had they been lying?' her daughter asked, somewhat bitterly.

The nature of light is a particularly vivid example of a topic for which more than one model is required for a full explanation. Another is the fundamental question of atomic structure. A model has to be presented early in school life to explain the basics of chemistry and physics. A simple orbital model, likening the atom to the solar system, is often given first, demonstrating clearly the separation of positive and negative charge. But a more sophisticated model, which explains atomic properties more accurately, associates the negative charge not with discrete particles but with a smeared-out cloud (Fig. 5.2).

To explain yet another set of observations – the way in which atoms interact with light – a third model is required. In this model the negative charges are not located at all, but are simply arranged in order, abstractly, according to their level of energy. Any attempt to visualise

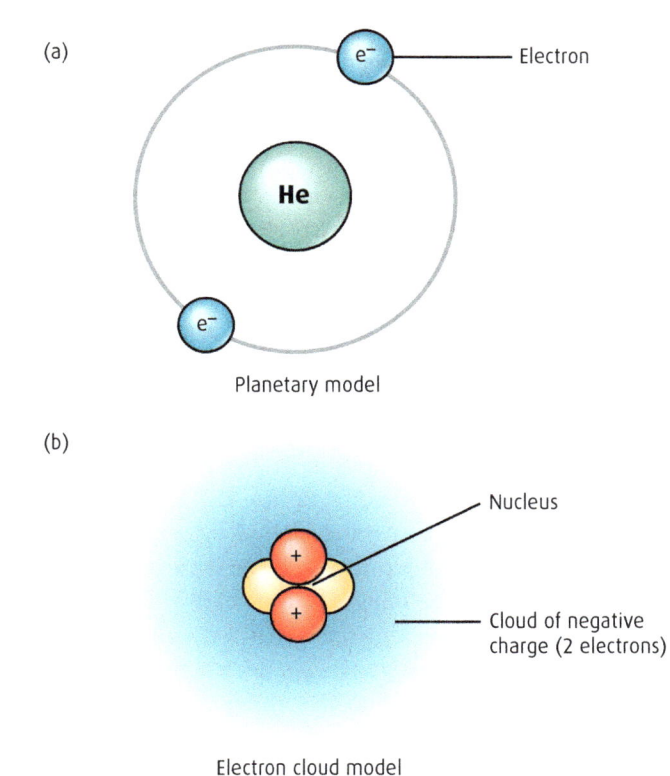

Fig. 5.2 Two distinct models of atomic structure

this situation has to be abandoned. Practising scientists may use any of these models, incomplete as each may be, in order to solve a practical problem. For people in one discussion group, adjusting to the idea of there being more than one model of a process seemed not only more realistic, but also rather comforting. It makes scientific understanding seem a little closer to general experience – complex and ambiguous, like so many other areas of life.

What is reality?

Armed with this new found freedom to employ models as required, what are we now to make of reality itself? What am I actually seeing when I 'see' a molecule or an atom? Photographs of atoms seen through electron microscopes are now widely available. They show images reconstructed by computer from the beam of electrons shone down on the specimen. Is this truly 'seeing'? What we actually look at is, of course, simply a computer screen which gives an image that has been constructed from the electrons bouncing off the specimen. Is this significantly different from a more direct act of seeing? When we see with the naked eye, isn't this simply a representation on the retina of light reflected off objects? Is seeing a computer image of an impossibly small atom – or indeed of a remote planet – much the same as seeing in the customary sense? Julie's earlier question to the particle physicist who described a blip in a stream of digits on a screen as 'seeing a particle' seems highly relevant.

In effect we have slipped quietly across the border from physics to metaphysics (from Greek, meaning 'beyond physics'). As human beings, whose early learning is based so largely on direct experience, we seem inclined to give particular credence to our visual sensations. We tend to believe things are real and actually exist provided we can see them. We can certainly see images of atoms, so are the atoms real? We certainly consider bacteria and viruses to be real even though they can't be seen without a powerful microscope. Is the light impinging on our retinas from these microscope images categorically different from that from a landscape or a face? 'Seeing' is not such a straightforward notion, after all. From physics and biology we can get a sense of what is actually taking place when light from objects enters our eyes and registers a nerve sensation with the brain, but it's more the metaphysics you'll have to consult to decide on whether atoms can be seen. Is the landscape real, but the image of an atom not? Or does reality cease to have a clear meaning at this point? We've moved beyond my competence; the choice is yours.

Fascination with colour, in paintings, in food, in sunsets led us to explore, in an earlier chapter, the nature of light, the carrier of colour. The quest seemed straightforward at first: light is so familiar, we are all bathed in it every day, all day. Yet as the story unfolded we were led ultimately to reconsider the very nature of reality. Light, it turns out, is indefinable in simple, unitary terms. It seems to behave as a wave in some situations and as a stream of particles in others. By attempting to explain the results of different experiments with light we are forced to question even what we mean by a model.

The effects of light beguile us as we perceive a setting sun or remote star, and even delude us when we respond to the images and illusions it creates nearer to home, in mirrors for example. But it would be misleading to attribute these effects to the tricks of light alone. Its inseparable accomplice is the seeing eye, the organ whose existence is dedicated to sensing the light falling ceaselessly upon it. The chapter that follows explores the eye itself and the way in which it responds to light, collaborating with the brain to create the experience of vision. Questions and observations from many different discussions are the inspiration for this chapter.

6
How We See

Illusion

It was not the anatomy of the cornea or the workings of the retina that had kicked off discussion of the eye in one group, but a talk Sonya had once heard about optical illusions, given by a scientist from the National Physical Laboratory. With the aid of her mobile phone she had shown the group images from the talk, revealing to everyone's surprise just how unreliable the eye's judgement can be.

In Fig. 6.1, for instance, the shades of grey at A and at B are precisely the same; it takes a second look to convince yourself (Fig. 6.1). As this test reveals, our judgement about the shade is influenced strongly by the immediate surroundings – in this case by the presence of a shadow cast by the cylinder over the B area. What does this bizarre observation tell us about the visual system? 'How can they be the same?' queried Sarah. 'You can see B is a light square and A a dark one.' 'Yes, but the vertical strip proves they are the same shade of grey,' Mary responded, grappling with the paradox. 'Surely the actual light entering your eye must be the same from both A and B if they are the same shade?' This was indeed a good point: the light intensity must be the same from each area, and therefore the amount of light falling on the retina at the back of the eye must also be the same. 'So the eye is registering the same shade of grey in each case then? I suppose it must be the brain that is doing the interpretation bit – working out there is a shadow so you imagine B lighter than it really is.' Mary had hit on a key concept: what we see is not purely a matter of optics, but the result of a team effort by eye and brain.

It turns out it is not only the shade on which the brain takes a view, but the colour and the perspective as well. At the same talk the speaker from the National Physical Laboratory had shown how our perception of colour changes when a particular colour is put alongside others.

Fig. 6.1 The brain interprets what the eye sees

Illusions can be created using geometric effects – cubes that appear to come towards you one minute then recede the next, for example. Or figures that appear as a vase in one viewing and as two faces in profile in another. We have seen in chapter 4 how the visual system is deceived in an even simpler way when it picks up images in a mirror. It can only understand light rays as having entered the pupil in a single straight line from an object. So rays from an object that have in reality bounced off the silvering of a mirror appear to the eye to have come from an image somewhere behind the mirror. Taken together, these various illusions reveal something about the visual system as a whole. Between the two of them, the eye and the brain work together to create a plausible impression of what is actually out there. Of course we rely utterly on this impression to navigate our way in the world, but these various 'tricks of the light' indicate something about the workings of the system behind the scenes.

The eye

'OK, Andrew, we get the idea, it's not as simple as just snapping a photograph. So how does it all work then? Light comes in through the pupil, we know that much. The lens focuses it on the retina...how the heck does it do that then?' In her customary direct way Julie had brought the

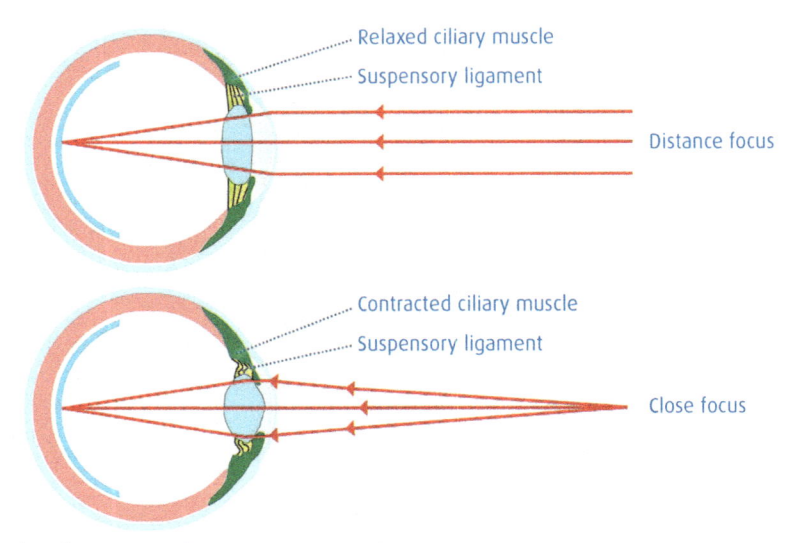

Fig. 6.2 How the eye creates an image

discussion round to a key question. How does the eye actually record the visual scene?

This diagram, perhaps familiar from a distant biology lesson, shows light rays passing through the opening of the pupil, on through the lens which then bends them to a focus on the back of the eye (Fig. 6.2). The muscles attached to the lens have an amazing capacity to adjust the thickness of the lens automatically, just the right amount to focus an object precisely on the retina, whether the rays come from nearby or far away. Of course, there's a greater degree of bending to be done if the object is close. That's why you can get an ache when you strain your lens muscle to look at something very close.

Unfortunately this system may not work perfectly for everyone. For some individuals the lens may be just too effective, bending the rays so much that they focus a little way short of their correct place on the retina. This so-called 'short-sightedness' is easily corrected using a lens that spreads out the rays a bit (a diverging or concave lens). The opposite holds for long-sightedness, remedied by a converging (or convex) lens. That's what spectacles do: reposition the focal point so it falls bang on the retina. No wonder it's important to get lenses of exactly the right bending power.

'Yes – I think I remember some of this from school biology,' interrupted Michelle. 'There are special muscles to change the shape of the lens, aren't there? But what I never got was how the retina works. Is it just a tiny screen at the back of the eye? What is it made of?' This image of a small cinema screen triggered a remark from Sarah. 'Detached retina... you get that, don't you? Is it like a screen that's got unhooked from its frame?'

This is where simple imagery begins to let you down, I am afraid. There is no little cinema at the back of the eye. The retina, like any biological material, is an intricate assembly of cells. Extraordinary as it may seem at first sight, the retina is in fact a continuous part of your brain tissue – part of the whole central nervous system. It's only half a millimetre thick, but in this narrow width lie several layers of different cells. One contains cells that have molecules in them that react to light. About seven million of these cells, known as 'cones', work in daylight and are sensitive to colour. Another group, called 'rods', are even more plentiful but work in low light, responding mainly in black and white.

'Amazing that so many cells are packed into such a small space!' said Julie. 'But when we say they "react to light" how does that actually work? Is it like photographic paper?' she added, always keen to probe a little further. 'Yes', joined in Mary, 'how on earth does something biological know there's light around? Surely it can't be electronic like a solar panel or chemical like a photographic film?'

In fact, when a particle of light (a photon) reaches the sensitive cell it is absorbed by a molecule in the cell, which as a result changes shape very slightly. Interestingly this molecule, called retinal, is a close relative of vitamin A, known to be important for good night vision. The diagram below represents this molecule before and after it changes shape (Fig. 6.3). The Cs, Hs and Os represent individual atoms, and other

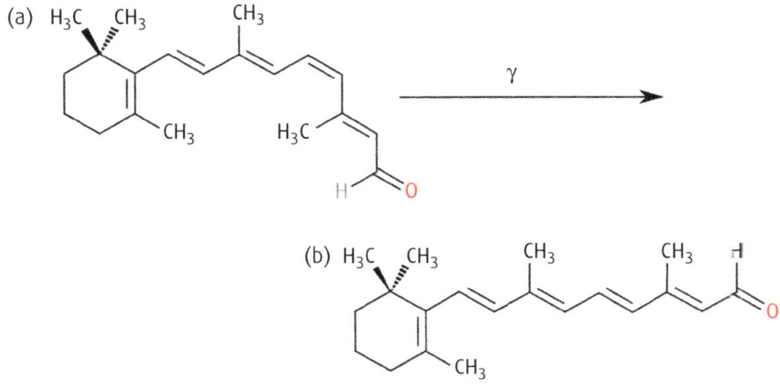

Fig. 6.3 Diagram of the retinal molecule (before and after a photon arrives)

unlabelled atoms are also implied wherever lines meet. Even without understanding any of the detail, you can see that it just takes a single shift in the arrangement of atoms to switch from the upper version of the molecule (a) to the lower version (b). This shift is caused by the arrival of a single particle of light (a photon, denoted γ in the diagram).

This physical change acts like a switch, triggering a series of linked actions that culminate in the cell changing its state of electrical charge. These light-sensitive cells are directly linked to other cells, a type of nerve cell called ganglion cells, which reach out beyond the retina and bundle together to form the optic nerve. It is along this visibly thick bundle of nerves that electrical signals flow directly from the eye to the brain whenever light falls on the retina.

'This is the point where I feel baffled,' interjected Mary, referring to her long-standing difficulty in grasping anything electrical. 'What I don't get is this: what do you actually mean by "a signal flows"? Are we saying it's like a wire? Are our bodies actually electrical, like a light bulb or a kettle?'

Nerves

Yes, a nerve is a bit like a wire – but only a bit. It's long and thin like a wire, and it carries an electrical signal from one end to the other. And yes, our bodies are electrical and the electricity in them is the same as you find everywhere else – there is only one kind. There the useful analogy comes to an end, however; the way electricity moves along a nerve is really quite different from the equivalent in a metal wire.

In a wire tiny particles called electrons, each one negatively charged, are moving around freely all the time. When a voltage is put across the wire from a battery or the mains it forces many billions of these electrons to drift towards one end (the positive end). This flow is called the current – by analogy with a river, I presume (see chapter 16). But, as you may have noticed, we don't have any long pieces of metal sitting in our bodies (unless you've broken a bone at some point), so there are no tiny electrons moving around freely. What there are however, in vast quantities, are various kinds of atoms that are electrically charged. These go under the name of 'ions' and are very commonplace things. For instance, whenever you dissolve common table salt in water or in the meat or vegetable juices on your plate you release ions – in this case ions of sodium and chlorine. These ions are simply atoms in which an electron normally attached to one atom (for example sodium) has shifted over to another atom (for example chlorine) sitting next to it.

'Why do these electrons move over like this?' you may well ask. It seems strange for an atom with a proper balance of positive and negative particles suddenly to lose one. It turns out that this happens because atoms achieve a lower state of energy when they contain a particular number of electrons (2 or 10 or 18 are examples). In this case sodium is energetically better off with one electron fewer and chlorine with one electron more And before Julie asks 'why is this so?', I'd better add that this is just one of the many bizarre rules that emerge from quantum theory. For now let's not go there, or we'll never get on to the brain. So dissolved in the fluids of your body are ions of sodium and chlorine from the salt you eat. There are several other types of ion too – calcium, potassium and iodine, to name just a few – you may have read about these on food labels. But what has all this talk of ions to do with nerve cells and how they send signals to the brain? Haven't we strayed from our path?

No, for it is indeed these ions that are key to how nerves work. An electron is electrically charged so, if it has left one atom to join another, each atom must have become electrically charged. One must become a little bit negative because it has an added electron, the other equally positive, because it has lost one. Therefore we can now see that ions are in effect electrically charged atoms, some positive, some negative.

'But what actually is a nerve?' inquired Michelle at this juncture. 'Is it a piece of flesh, made up of cells like other organs? How does it all connect up? How does it actually work?'

The word 'nerve' refers to a bundle of nerve fibres running side by side. A fibre is the long, thin part of a nerve cell known as an axon. Nerve cells (or neurons) are very unusual cells. Like any other cell they are made up of a surrounding membrane enclosing a nucleus and many other internal bits, but, exceptionally, one part of the neuron is grossly extended in one particular direction. The cell has a broadly globular body shape overall, but one part of the membrane runs outwards like a very long thin cable. It is this so-called axon that acts in place of a wire. It extends a long, long way – in some cases centimetres or even a metre – and carries the electrical signal throughout the length of the body. The schematic diagram below shows the various parts (Fig. 6.4). In reality these might be larger, smaller, thinner or fatter than those illustrated.

As this diagram clearly demonstrates, the neuron has not only a long extension or axon to carry the electrical signal, but also plenty

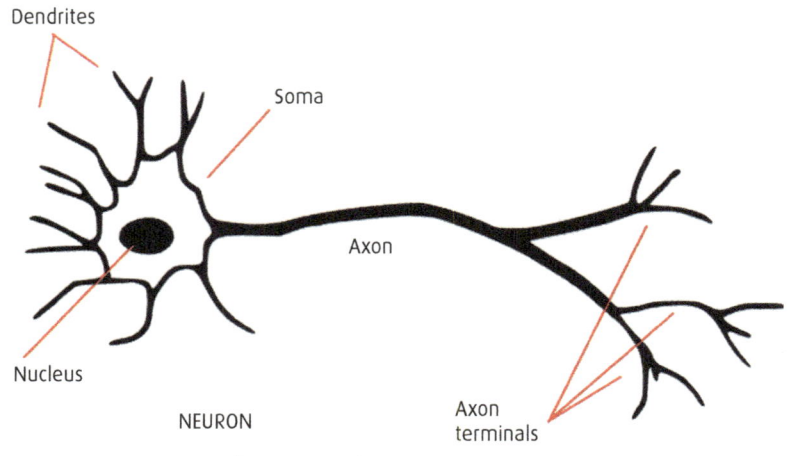

Fig. 6.4 Diagram of a nerve cell (or neuron)

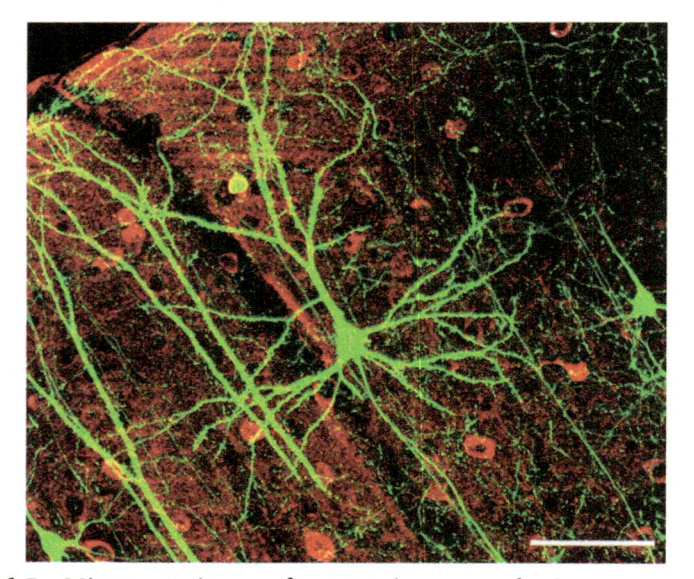

Fig. 6.5 Microscope image of neurons in a mouse brain

of branches at each end – known as dendrites and axon terminals. It is these vital terminals that enable the cell to pick up its message at one end and deliver it at the other –vital aspects of any signalling system.

The photograph shown here is not schematic (Fig. 6.5). It portrays an actual retinal cell, showing both the central body of the cell (or soma) and lots of dendrites branching off it that pick up signals from neighbouring cells. A single long, thin axon, extending from the lower right of the central body, becomes a strand of the optic nerve leading to the brain.

So that's the basic structure, but what about Michelle's main question: how does it all work? The membranes of nerve cells, their outer 'walls', have the remarkable ability to sustain a voltage difference. Between the outside and inside surface of a membrane there is a voltage difference of about one-fiftieth of a volt – a very small and harmless amount. When a cell is stimulated, often by a chemical trigger from a neighbouring cell, this voltage suddenly rises (or falls), making a short pulse. This effect happens as a result of electrically charged ions moving successively and very rapidly in or out of the cell across the membrane that surrounds it. An amazing consequence follows this simple and momentary pulse. Rather like a Mexican wave rippling round a football stadium, as one person after another stands and sits, the electrical pulse moves away from its starting point and passes along the cell membrane. This is explained in greater detail in the next chapter (see Fig. 7.2).

Precisely because a nerve cell has a specially adapted long, cable-like 'axon', the pulse can travel a very long way down it indeed. In the case of the optic nerve the axon runs all the way to the brain, carrying electrical pulses from the retina of the eye to the visual part of the brain. In essence, this is how the things that stand before your eyes are received as impressions on the brain. Electrical signals generated by light falling on the retina end up as electrical pulses flashing along the optic nerve, so sending information to the brain. This is the basis for the images we create of the world around us.

The brain

So 'seeing' is really a double-act between eye and brain. The eye plays its part by registering the light falling on it from objects around, creating corresponding electrical pulses and delivering them to the brain for processing. The obvious next question was put by Mary. 'How does the brain take it from there?' Answers to this question are now beginning to emerge because in recent decades research on the brain has simply raced ahead. This is partly due to the steady march of experimental psychology and neurology, but more recently there's also been the dramatic development of imaging technology such as the MRI scanner. This has enabled associations to be made between regions of the brain and specific functions (see chapter 17, MRI and the Brain). In the case of the visual system, two distinct areas of the brain seem to play important parts. There is an intermediate area where the first stage of processing occurs and a region at the back of the head, called the visual cortex, where it continues.

Understanding the workings of the visual cortex is a highly active area of current research. This means, of course, that we don't yet know all the answers. Some studies involve presenting particular visual images to the eye and studying which neurons seem to respond. Others involve MRI scanning to see which areas of the brain are active under various different conditions. A third type involves studying the effects of various visual deficiencies presented by patients. The picture is far from complete, and several theories have been put forward to explain the visual cortex. What they have in common is the idea that a huge variety of different neurons is involved, each specialised in one way or another. Some detect movement, some colour; others respond to edges in the visual field, and yet others to whole forms such as the human face. Different sub-sections of the brain tissue in the visual cortex contain different amounts of each type of neuron, and so are themselves specialised for particular functions, such as detecting movement.

But this increasingly technical talk of the anatomy and function of the brain puts us in danger of straying from the original question. As Julie said at this point in the discussion, 'Can we remind ourselves why we are here? Weren't we basically interested in how we see things? We started with finding out what light is; that took us to the eye and then the brain.' 'I can see the brain is essential to understanding how we see things,' responded Mary, 'but what really intrigues me is not just the mechanics of how it all works, but how it relates to the psychology – what you feel about what you see.' 'Yes, yes,' agreed Sarah. 'I remember hearing about an amazing experiment years ago with a bear or something running right across a screen and nobody noticing it. What was that all about?' Her memory was correct: a psychology experiment had once involved asking people to watch a video of two teams passing a ball and to count the number of passes one of the teams makes. After having concentrated intently on the counting task for a minute or two the subjects were surprised to find they were asked not about how many passes they had seen, but whether they had noticed the bear. Looking baffled, most said no, but on reviewing the video they were amazed to see that an actor dressed as a bear had in fact walked nonchalantly right across the field of view – plain as a pikestaff. By focusing on one visual task they had inadvertently created a frame of reference that excluded irrelevant visual material.

The point raised by Sarah was that there is more to what we perceive than what is in the visual field. The brain makes choices about what to focus on, what matters most in any given context. Mary had raised a further point about the psychology of perception: it's not only a matter of *what* your brain allows you to perceive in the visual field, but also how it makes you feel. Some impressions make us feel pleased – a beautiful landscape, a smiling baby – while others fill us with fear or disgust – a large hairy spider or a rotting carcass, for example. 'How does the brain affect what we feel?' was her question. 'How does some electrical impulse circulating round the neurons of the brain lead to me feeling happy or sad?' In raising this new and intriguing set of questions she had inadvertently moved us on to another stage of this unfolding story, to be discussed in chapter 7.

The brain plays a crucial role in how we see the world, interpreting the visual messages arriving along the optic nerve from the retina. Trying to work out how the brain works and what it means to 'interpret' messages is now a burgeoning field of scientific research. Not surprisingly the science of the brain is a regular topic in discussion groups, with their plentiful questions about everyday life and its meaning. Sometimes it is mental health that provides the trigger, sometimes an issue of child-raising and human development; from time to time it's the deeper questions of feeling, thought and consciousness that drive discussion.

In the next chapter, the broad field of brain studies is the subject and the questions asked by people are the guide. What follows is not a complete description of a complex and rapidly developing field. Rather it is an introduction to some of the themes that arise when inquiries are made by ordinary people following up their curiosity.

7
The Brain

'It's extraordinary to think that the brain, with all its thoughts and feelings, is driven by chemicals and electricity.' Helen captured in a sentence what most members of the group had been silently thinking. Yes, it's fascinating to hear about the latest MRI research and read about the various parts of the brain with their complicated Latin names, but at the fundamental level it's really hard to imagine our mental experiences happening in this way. Neurons flashing on and off, endorphins flushing through the grey matter: it all seems disconnected from the everyday thoughts and dreams that occupy our minds. Of course people have different beliefs about the ultimate basis for our emotions and reasoning – but whatever these may be, an increasing body of facts about the workings of the brain are being uncovered by fast-moving scientific advances.

Individuals in the group had read various articles and books linked to the subject, and some had watched documentaries on television. More directly, issues of mental functioning and personal development had touched the lives of everyone in the group at some point, in one way or another. No-one had particular expertise in brain science, but all had reasons to be curious about it. As a consequence the group decided to settle on the subject for a few months, to pool their experiences and see how far a little reading and surfing would take them.

Pooling experience

Sally kicked off an exchange of experiences with the story of her niece who had suffered a brain tumour at the age of four. Surgeons had operated to remove it and a surprising consequence was that her personality had changed dramatically. From being a relatively shy child she became overnight much more outgoing. Helen recalled a television programme in which damage to a person's brain tissue had resulted in her becoming addicted to gambling. As Rosie observed, 'it seems strange that an addictive tendency could be due to a part of the brain rather than some kind of external influence. You'd think gambling would take a hold because of the lure of financial gain or the influence of friends. I suppose it's easier to imagine a chemical influence in the case of alcohol or drugs.'

Almost everyone round the table had experience of a friend or relative with a mental health problem, depression in particular. Sally and Patrick both knew people who had been treated with ECT (electroconvulsive therapy); it had fortunately proved beneficial in both cases. 'But what is ECT exactly?' asked Rosie. 'It seems a bit brutal to simply shock the brain indiscriminately.' She answered her own question shortly afterwards by consulting the Royal College of Psychiatrists website. Apparently the electric shock induces a kind of epileptic fit. It had been noticed that people with epilepsy seem to feel better after having a fit. It is not fully understood how this happens, but it is known that depression is associated with altered behaviour of certain chemicals in specific parts of the brain. It is thought that the epileptic fit may influence this for the better. Given the risk of side-effects, on memory for example, the treatment today tends to be restricted to severe cases for which other treatments have failed.

A malfunctioning brain seemed to be at the heart of other problems people had heard about. Alzheimer's was affecting the elderly relatives of several people round the table. Julie's mother, for example, had been gradually losing her memory. 'It's bizarre and unpredictable. She can write her name and her daughter's, but not her granddaughter's.' Others mentioned the mental health of children and teenagers, focusing on autism as a condition that has become better known through the book *The Curious Incident of the Dog in the Night-time*. This is another example of a specific brain disorder about which awareness is increasing. It affects information processing by altering how nerve cells in the brain connect up, though the details of how this happens are not understood.

As discussion progressed, it became clear that the group was tending to focus on disorders of the brain. Mental health seems to be increasingly talked about today – a good thing and a marked improvement on the past. Perhaps it was the changing culture that was causing this: a decreasing sense of stigma, less covering up of conditions. Or was it the result of better treatments, or increased understanding brought about through the advent of new technologies such as the MRI scanner? Whatever the cause, everyone seems to be acquainted with the effects of mental ill-health in one way or another. But, as Rosie pointed out at this point in the discussion, 'Shouldn't we also be looking at the brain from a positive point of view? After all, it is responsible for the way we think and feel in good times as well as bad!' 'Of course,' others chimed in, 'it's the brain that does our thinking. It's creative, imaginative and full of thoughts. It's responsible for our feelings of happiness and joy as well. Where would we be without these?'

So how do we begin to discover more about how it all works? Perhaps one starting point is to map out something of the various branches of science that contribute to our understanding of the brain. After all, it's not only brain surgeons who know something about the brain.

Studying the brain

An arcane subject just a few decades ago, with little written for the general public, brain science has been become more widely recognised recently through the popularity of accessible books such as Oliver Sacks' *The Man Who Mistook his Wife for a Hat*. Much of the evidence for the stories of people with extraordinary cognitive disorders came originally from neurosurgery – in particular from case histories of people who had suffered injuries to specific parts of the brain. Patrick had read about an example in which the tissue that connected the left and right halves of a person's brain had been cut in an operation, pretty much a last resort intervention in a case of severe epilepsy. When a cup was placed in one of the patient's hands he had no problem recognising it; when placed in the other hand, he could feel the object but not name it. This sad example provided important clues about specialisation in the separate halves of the brain. It is now understood that the right hemisphere is involved in spatial tasks and with emotions such as empathy, humour and depression, while the left is more dominant in verbal tasks such as speaking and writing and for scientific and mathematical skills. However, popular myths about left-brain and right-brain personalities are rejected by neuroscientists – now actively researching the complex ways in which the two halves interact and, to some extent, overlap with each other.

A quite different area of research about the brain is the field of experimental psychology. These studies are not aimed so much at explaining precise mechanisms of the brain, but at measuring what happens under varying experimental conditions. In this way properties of the brain can be inferred. An interesting example was raised by Patrick who had been reading the popular science book *How the Mind Works* by Steven Pinker. A psychologist, Paul Ekman had researched whether facial expressions of emotion developed culturally as a baby grows up or were universal. It appears that many expressions, including anger, disgust and fear, are, to a large extent, the same in all cultures. Other branches of experimental psychology focus on our senses, the portals through which the external world impacts on the brain. Research into visual and aural perception, for example, reveals ways in which the brain interprets the world. The way we perceive certain images, for instance, has led to the concept of Gestalt: the idea of an overall form which the brain imposes on elements in a sensory field. The theory explains how the perceptual system forms a percept of the whole as a reality on its own, independent of the elements of which it is composed, as the following diagrams illustrate beautifully (Fig. 7.1). When you gaze at these figures, the apparent white triangle in (a), for example. is created in the brain; so is the sphere in (c).

Early studies of the brain focused, as you might expect, on its physical make-up, its anatomy. Studies of the structure of brains of the deceased enabled many distinct parts to be identified. Links between these structures and a number of functions were established by studying people who had suffered damage to specific regions. For example, the

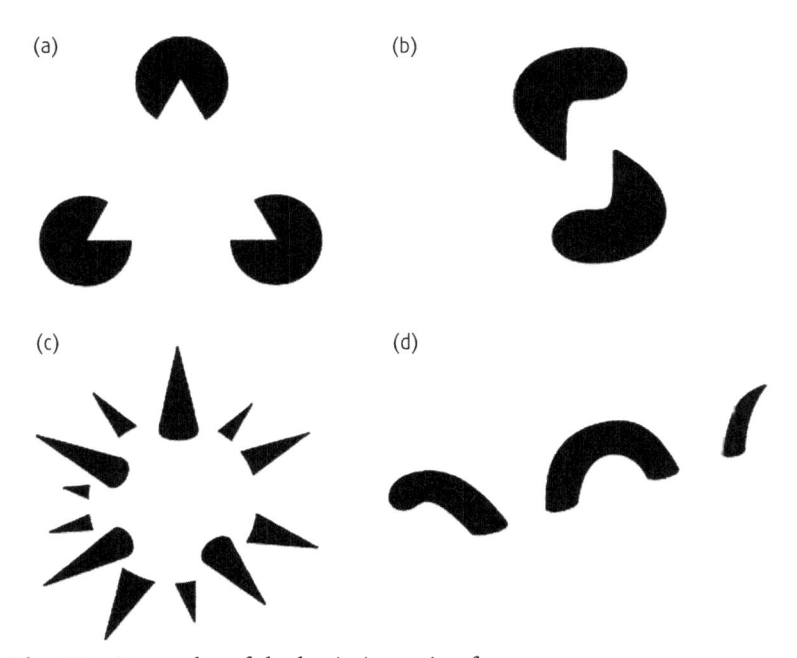

Fig. 7.1 Examples of the brain imposing form

brain of an individual able to understand speech but not to talk was analysed *post mortem* by the French anatomist Paul Broca. He found damage in a particular area at the front of the brain. Such studies demonstrated what we now take for granted: distinct regions of the brain are associated with particular aspects of our functioning and behaviour. We now know, for example, that a region at the back is important in visual perception. The frontal region is associated with thinking and planning, among other things, and a deep region in the oldest part of the brain, in evolutionary terms, controls our breathing and heart rate.

This concept, known as localisation, has developed enormously with the advent of modern scanning technology. 'Ah yes,' Sarah chipped in eagerly at this point in a discussion. 'My aunt had an MRI scan. What does this actually do? It's a pretty noisy procedure, does it do any harm?' MRI machines are relatively new devices that use magnetic properties of the atoms in your body. Unlike when you have an X-ray, no radiation is shone on to your body. Instead strong magnetic fields are applied; these affect the hydrogen atoms in the water of which much of your body is composed. This causes the atoms to emit electromagnetic signals that are picked up by the machine. The scanner is able to pinpoint exactly where the signals are coming from, enabling it to create a remarkably precise map of the body. See chapter 17 for more detail.

In brain studies this technique is applied in what is called functional MRI scanning to identify places where blood flow is increasing. This occurs wherever neurons are active, thus revealing areas of the brain that are functioning at any given moment. In this way the parts of the brain that are active under different circumstances – thinking, seeing or hearing, for example – can be identified. Through Sarah's question another group of disciplines that contribute to modern neuroscience had been identified – the engineering and basic sciences that are developing the technology for neurologists and anatomists to use.

As the discussion group gradually became aware of the basic geography of the brain, a fundamental point became increasingly clear. The human brain was quite obviously not created, as it were, in one go; it evolved over time, as *Homo sapiens* itself evolved. It is not as though the ideal brain design just appeared. The brain we have today is an accumulation of older and younger parts, reflected in their physical locations and in the type of function for which they are responsible. Thus the oldest part, called the brain stem, is right at the bottom of the brain where it becomes the spinal cord. It plays a key role in fundamental functions: sleeping and eating, breathing and maintaining heartbeats. The most recent part, the large cerebral cortex, lies higher up on the top of the brain and is associated with higher functions, including language and thinking. Interestingly all vertebrates share a common basic form, a three part system of hindbrain, midbrain and forebrain. The brains of mammals, one particular class of vertebrates, are distinguished by being generally much bigger, with a more developed cerebral cortex.

How does it all work?

The layout of the various parts of the human brain and the way in which they have evolved are central to the descriptions of the organ given in books and websites about human biology. However, these are rarely the starting point for people in discussion groups who want to know more about the subject. Their concerns are more related to how the brain affects our daily lives. 'Isn't memory bizarre?' exclaimed Helen on one occasion. 'You can forget the name of someone familiar to you, yet recall the exact details of some music or smell from years ago.' 'Yes,' added Sally, 'what is happening when you forget something and are reminded of it? Sometimes you recall it, other times you don't.' 'And what about false memory – what on earth is going on in the brain there?' asked Rosie, thinking of a recent legal case. 'Does the brain contain the memory or not?'

Memory is indeed a popular area of discussion. Another is the social aspect of brain functioning – the effect of the early environment on infants' brain development and the way in which the brain affects mood and behaviour through its interaction with hormones. In the remainder of this chapter we follow these matters of interest, but first a word or two about the physiology of how the brain works – the basic processes.

Neurons

The fundamental unit of the brain, and indeed of the whole nervous system, is the nerve cell or neuron. Its structure is outlined briefly in chapter 6. Here we look a little more closely at how it functions. A typical human brain contains an unimaginable number of these cells, approximately 86 billion. This huge number is comparable to the number of stars in the Milky Way galaxy, estimated at between 100 and 400 billion. Difficult though these numbers are to grasp, the really important figure for understanding brain capacity is not in fact the number of neurons, but the number of connections they make with each other at junctions known as synapses: 100 trillion, roughly one thousand times the number of neurons. This tells us something important about brain functioning: each neuron must have many, many synapses – places where one neuron connects to another. It's the degree of connectivity, not the number of 'wires', that really counts. So the brain is a kind of electrical signalling system in which each component is linked to up to 1,000 others. This is what distinguishes the brain – not its size nor the number of its cells, but their connectedness.

Synapses

Neurons are specialist cells, characterised by their unusual long, thin extensions (called axons) along which electrical pulses pass. The dimensions of these axons and their electrical properties are reminiscent of electrical wires, but, as indicated in chapter 6, electric pulses travel along them in a quite different way. Rather than the steady flow of electrical charge found in wire circuits, momentary pulses of rising and falling voltage pass along the membranes of nerve cells. They do this, when stimulated, by allowing electrically charged atoms (known as ions) to flow in and out of the membrane of the cell in a coordinated sequence, rather like the up and down movement of individuals in a Mexican wave (Fig. 7.2).

In this way an electrical pulse moves along the length of a long axon right to its tip. Linked to each of these axons are a number of connections to neighbouring nerve cells. It is at these junctions, the synapses, that the signal from one neuron is passed to the next. But this connection process is quite unlike the usual electrical procedure of linking two wires together, in which the charge flows continuously across from one site to the next. The biological equivalent is a more complicated process; it is central to the way we think and act, and is implicated in the onset of brain disorders such as Parkinson's disease. At the junction between two neurons, the pulse travelling down the first one triggers off the release of a chemical substance, which passes out of the first neuron, travels across the tiny gap to the next neuron and passes through the membrane of the second neuron. Inside the second neuron this chemical, known as a neurotransmitter, initiates a new electrical pulse which, in effect, carries the signal forward. In this way a message is passed by relay from neuron to neuron, like the lighting of a chain of beacons (though somewhat faster!). Armed with this vivid image of messages travelling along nerve cells jumping from one to the next, thanks to the neurotransmitters, new questions naturally arose in the discussion group. 'So are these

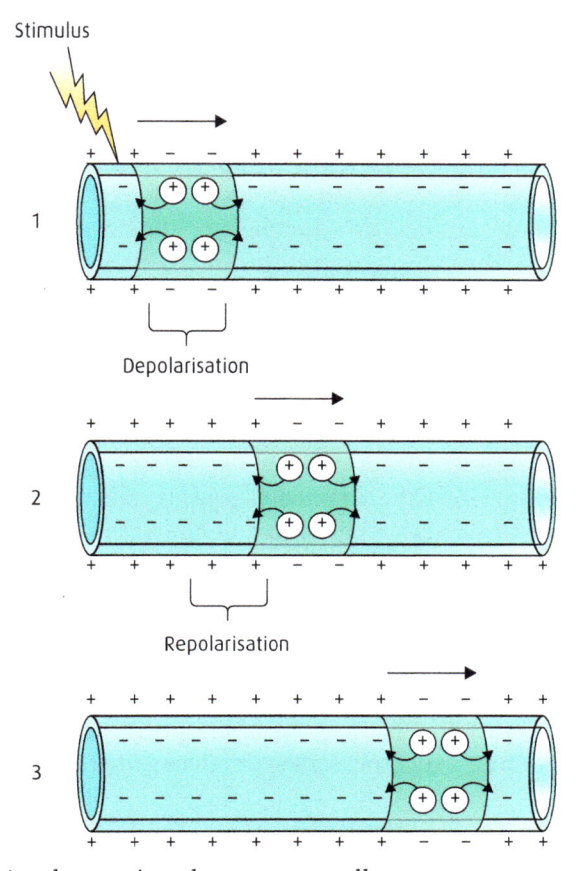

Fig. 7.2 A pulse moving along a nerve cell

the chemicals in the brain people talk about?' Helen asked. 'Is this where dopamine comes in? Isn't it involved in Parkinson's disease?'

Yes, dopamine is indeed one of the neurotransmitters. Surprisingly there are quite a few different ones. You might have thought one chemical would be enough to link up different nerves, but evolution has left the human brain with many – dopamine, serotonin, acetylcholine, histamine, adrenalin, endorphin, to name but a few. In fact there are many different types of neuron mediating different functions in different parts of the brain. Different neurotransmitters are associated with each of these. 'OK, so let's take Parkinson's disease as an example – what's the role of dopamine?' asked Rosie. 'Isn't it part of the treatment?'

The symptoms of Parkinson's disease are caused by a decrease in the levels of dopamine, due to the death of the cells in the brain that make the substance. In recent years research in biochemistry, physiology and pharmacology has enabled drugs to be developed that mimic dopamine and thereby stimulate nerve cells. Though not the cure we all wish for, such drugs can help to alleviate symptoms. Developing treatments of this kind is an important motive for research in the basic as well as the applied sciences. Even though fundamental research of this kind is essentially exploratory – and therefore its outcome is unpredictable – understanding of basic structures and mechanisms is essential if treatments are to be developed.

One discussion group gained direct insight into this kind of research by arranging a visit to a researcher in a cell biology laboratory. Invited to peer down a microscope at a mass of tiny nematode worms, just a millimetre long and wriggling around full of life, the group heard these were essential to fundamental research into depression. Apparently, despite the relatively tiny size of the nematode's nervous system, the mechanism of nerve transmission is similar to that of humans in important ways. The neurotransmitter serotonin is present in both, and the way in which it transmits signals can be explored more easily in the simpler species. Low levels of serotonin are associated with depression in humans and a comparable effect can be observed in nematodes: they stop wriggling about. This example illustrates how an area of fundamental research, in which a simple creature functions as a laboratory model, can play such a vital role in informing the subsequent development of drugs for human disorders.

In recent decades a very important discovery has been made about the way in which synapses, the junctions between nerve cells, are strengthened or weakened through experience. It turns out that each time a synapse is activated it not only transmits a signal to the next cell; it also consolidates its own existence. The more times a synapse gets used, the more secure is the connection it makes. This phenomenon, known as synaptic plasticity, has remarkable implications. Researchers are now beginning to develop a model of the biological mechanisms of memory and learning. We all know that learning necessarily involves a degree of practice and repetition; now we see there is a corresponding process of repetition at the cellular level. Networks of neurons become strengthened in the brain when the connections between neurons are repeatedly activated. In addition to the implications of this 'plasticity' for memory and learning, it also modifies our ideas of the brain's capability. Far from being a fixture, as previously thought, we now know that brain capacity is capable of development throughout the course of life. This new understanding is already affecting the ways in which stroke patients and others with nervous system disorders can be helped. It also adds to the case for adult education, reinforcing the concept of lifelong learning.

Brain and hormones

At this point discussion might well have deepened, to look into the mechanism of nerve transmission and its defects. After all, many important ills including addiction, autism and depression, are associated with defects at the synapse. As it happened, however, the conversation turned in a different direction altogether: to the link between the brain and hormones.

Stephanie, an experienced psychotherapist, took up the thread. 'Some of the chemicals in the brain we have mentioned affect our moods, don't they? Take adrenalin, for example: as far as I know it surges when you are in danger. Doesn't it cause the "fight or flight" response?' 'Good point,' agreed Helen. 'I have often wondered: if the roof falls in, how do I know how to act? Would an animal know there was danger if the roof falls in?' This point was taken up by Sally, who observed that if the roof collapses your brain picks up signals immediately through the visual and auditory senses – and presumably many kinds of animal would react in a similar way. Rosie added that learning must also be involved; 'after all a child learns that fire is hot'. 'Animals have built-in reactions as well as learned ones,' she continued. 'Think of birds; they use some kind of magnetic sense to navigate as they migrate, don't they?'

A more philosophical note was introduced when Rosie raised the issue of the division of mind and body. 'In Eastern philosophical traditions,' she said, 'they just aren't treated as separate.' 'Isn't it down to Descartes that we in the West think of the two as so disconnected?' Helen responded. 'It makes it difficult for us to reconcile experimental evidence about the brain with our personal experience of living.' This latter point neatly captured an undercurrent running throughout the entire discussion. In effect an objective scientific view, based on emerging research and incomplete information, was being combined with subjective points of view based on people's direct experience of life and love. With the traditional separation between mind and body, we are constantly busy trying to figure out what is a cause and what an effect. Did the adrenalin start coursing through my veins because of the fear? Or did its presence in my blood cause me to feel frightened? Which caused me to take that mood enhancing drug: me with my freely chosen desires or my neuron circuits with their addictive response? 'At least in some respects we are getting better at understanding this these days,' commented Sally. 'At least we have the concept of psychosomatic causes which is generally accepted. For instance, my GP told me that she is doing research on how the occurrence of accidents seems to be related to the levels of stress a person is experiencing.' The relationship between hormones and the brain took hold of the group, and to investigate further the members decided to read around the subject. A particular feature was found that provided a closer look at how the two interact: the pituitary gland, buried deep at the base of the brain.

The case of the pituitary gland

The pituitary gland is a kind of super-gland located at the bottom of the brain. What first struck the discussion group was the extraordinary contrast between its size and its importance. 'It's amazing!' Helen exclaimed. 'How can such a small thing be so crucial to our hormone balance?' The pituitary is often referred to as the 'master gland' because it controls several others – the adrenals and thyroid, for example – yet it is only the size of a pea. It sits in a bony hollow below the base of the brain, behind the bridge of the nose.

The great interest of the group, however, was less in the anatomical detail and more in how events in the brain affect our feelings – how nerves interact with hormones, in other words. How does a chemical affect our behaviour? What makes a child grow so rapidly? Why do children turn grumpy in their teens? What triggers animals to spawn and to migrate? A host of interesting questions followed, pointing to the link between the brain and the endocrine system – the complex system of hormones.

We have seen already how messages get relayed around the brain through complex networks of nerve cells, linked together in their billions. We've also seen that this biological network is not organised in quite the same way as a network of wires that you might find inside a computer, for example. As we saw, it's not that electric current flows from one wire to other wires connected to it. Instead neurotransmitters are released which pass out of the cell and diffuse across the gap to the connected cell where they trigger off a new pulse. So much for a reminder of how the synapse works; now what about the pituitary gland? What happens here?

It seems that a somewhat similar process happens in the gland as in the nerve. But in this case it's not that an electrical signal in one nerve cell kicks off another one in the next cell; instead an electrical signal kicks off a chemical signal that goes on to affect the whole body. In effect electrical pulses from nerve cells in the brain cause hormone molecules to be released into the bloodstream. These hormones then spread all over the body, producing their various effects. The details of this process have been revealed by painstaking research in many disciplines over recent decades. It turns out that neurons in the brain are able to pass their electrical signals to a particular part of the brain called the hypothalamus, located at the base of the brain. In this specialised region the neurons (called 'neurosecretory cells') have a special property. At the far tip of their long axons these cells contain small packages which are filled up with hormone molecules. But these specialised neurons are not connected to further neurons, as in most parts of the brain. Instead they connect directly to the walls of a neighbouring blood vessel – a tiny capillary. By passing through the permeable walls of the capillary these hormone molecules are released into the bloodstream, ready to circulate freely throughout the body. 'It's as though this is the moment at which mind meets body,' said Jean, thrilled by the idea of a link between electrical signals flashing through the brain and hormones flowing in the blood.

The hormone exerts its effects in a two-stage process that seems unnecessarily complicated at first sight. The job of this hormone is to stimulate further types of hormone which then circulate around the body. In this two-stage process the original stimulating hormone released from the nerve cell goes on to release a variety of other important hormones. These include growth hormone, puberty hormones (gonadotrophins) and ACTH, a hormone which itself goes on to stimulate yet another hormone, cortisol, strongly associated with stress. In each case the action takes place well away from the pituitary, in whichever part of the body is appropriate – the kidneys or gonads, for example. But, as Anna suggested, even though there are a limited number of types of hormone it's reasonable to expect a wide range of different effects. As she put it, 'cake recipes only involve a few ingredients, but there are endless variations in what you can make with them.'

Stephanie, the psychotherapist in the group, found this insight into how the brain influences our emotions particularly interesting. In her professional life she deals with children and young adults living with exceptionally high levels of stress. In her understanding it was quite possible for trauma experiences early in infancy, especially if repeated, to set the stress response at the wrong level; this can go on to affect responses in adult life. This observation chimed with a common feature of hormone systems. Their job is often not just to set a process in motion, such as lactation or bone growth, but also to regulate it, so that it operates at the right level, neither too much nor too little. In this respect hormone systems are analogous to a thermostat which regulates the temperature of a room by taking action if it falls too far or the opposite action if it rises too high. Regulating body temperature is an example of this, and ensuring the right level of sugar in the blood is another.

Conclusion

These insights into the nature and role of hormones inevitably throw up new questions. As you might imagine, the discussion group was not slow to pose a stream of them. What do hormones look like? Are they all the same kind of thing? Where do they come from? If they are just messengers how do they produce their effects? How do they know when to switch on for puberty and pregnancy? Most fundamental of all: what exactly is a hormone? Eager as the group was to pursue these questions, a detectable glazing of their eyes spoke clearly: there's only so much you can take in at a single sitting. The hidden world of hormones became the subject for many further months of discussion and as a consequence, forms the basis of the next chapter.

Before moving on to this intriguing subject, however, let's reflect briefly on the many ideas about the brain presented in this chapter. We've seen how research in many disciplines – surgery, anatomy, physiology and psychology, for example, to say nothing of physics, computer science and philosophy – has combined to give us our present insights. From the nineteenth century concept of phrenology to contemporary images derived from magnetic resonance technology, the geography of the brain has provided vital knowledge. So too have detailed studies of the structure of individual nerve cells and the chemistry and physics of their operation. We've explored the all-important synapses, the junctions between nerve cells, and the networks they give rise to. Finally, in a move to connect our 'thinking' brains with our 'feeling' bodies, we have seen how nerve cells are able to unleash hormones into the bloodstream, producing the extraordinary changes we all experience as we grow older and grapple with the complexities of life.

The next issue to tackle is how the firing of cells up in the brain can give rise to physical and emotional changes throughout the rest of the body. How does a stimulus from the eye or a long-stored memory set the heartbeat racing or unleash a sense of joy? One particularly important process for this involves the complex system of hormones. Linked as they are with feelings of wellbeing or sadness, and with profound changes affecting our bodies at crucial stages of life, hormones figure prominently in group discussions. What follows is an account of several such discussions and some of the research that accompanied them.

8
Hormones

'It can feel as though a huge syringe full of heat has been put into you and shoots through your body. It makes you sweat, gives you clammy skin'. So declared Julie in a vivid description of a hot flush, as regularly experienced by women at the menopause, but not so often expressed openly. 'It seems to alter your body's control of temperature,' she added, slanting her comments towards the scientific aspects of the sensation for the benefit of the discussion group. 'Does your body actually get hotter?' she continued. 'Come to think of it, how is its temperature controlled anyway?'

'I think it's something to do with hormones, isn't it?' responded Sarah, who had once been to a talk on the topic. 'I think I remember there has to be a balance of oestrogen and progesterone, or something like that,' she went on, struggling to recover a distant memory. A quick search on the internet showed that hot flushes are caused by a reduction in the levels of oestrogen and progesterone produced by the ovaries. We are in the realm of hormones, which became the subject of the subsequent discussion and indeed of this chapter.

A subsequent internet search for the causes of hot flushes led to an interesting 2012 study using rats by Dr Naomi Rance at the University of Arizona. Apparently a particular group of neurons, or nerve cells in the brain, plays an important role in how the body controls both temperature and reproduction. These cells are in the hypothalamus, an area we encountered in chapter 7, which is the crucial meeting point between electrical signals in the brain and the hormones that mediate physical actions in the body. Using rats, researchers were able to investigate the biological mechanisms of the menopause. It turns out the trigger is a reduction in the level of the hormone oestrogen. The temperature changes because blood vessels are widened, which increases blood flow through the skin. This appears to be an effort by the body to get rid of heat. In fact the temperature deeper inside the body is not affected at all; it stays the same. 'Very interesting. Clearly a very intricate mechanism at work, but what is it that begins to happen to women at a certain age?' asked Julie.

According to the Arizona research team, it looks as though the particular brain cells in the hypothalamus mentioned earlier are sensitive to oestrogen. At the same time, however, they are connected to other parts of the brain that regulate body temperature. When levels of oestrogen fall too low, these cells in the hypothalamus start to malfunction: they begin to send out inappropriately strong signals. Unfortunately for the person involved, these signals get picked up as instructions for the body to lose heat quickly – hence the widening of blood vessels close to the skin.

'So it's all a mistake in the body – there's no reason for the heating up?' interjected Sarah, somewhat surprised. 'Is it just bad luck that the part of the brain that regulates temperature happens to be closely linked to the part that controls reproduction? Why are women's bodies designed in this way?' asked Julie jokingly, knowing full well that nature is not intentional in this way! Bad design is not quite how researchers see it, of course. There may be an evolutionary explanation of how regulation of temperature and reproduction became linked. Perhaps, they surmise, it's because hypothermia is such a risk during pregnancy. The slow pace of evolution means our bodies remain adapted to the risks of early human life millions of years ago, rather than to the world in which live today.

So for the present, as so many of us live on to an age at which oestrogen levels begin to diminish, the only 'remedy' is to replenish the missing hormone – hence the common treatment: hormone replacement therapy. The hope of researchers is that a new preventative treatment will ultimately be developed as a result of understanding more precisely the mechanism that causes the flushes.

What are hormones?

This talk of hormone levels and control mechanisms in our bodies raised a flurry of further questions in the group. What exactly is a hormone? Sharing first thoughts about this fundamental question revealed just how important and indeed influential they seem to be. 'Where do we begin?' Sonya responded smiling, fresh as she was from a training course in body therapy. 'What about oxytocin, which I have been told plays a big role in soothing people? Apparently it's released when you stroke people.' 'And of course there's the hormone in "the Pill", oestrogen, which must be mixed up somehow in the reproductive cycle,' commented Sarah. 'And there's testosterone too,' Patrick agreed. Mary took another tack, recalling tentatively that 'her aunt had had a problem with her thyroid – I think she felt excessively tired, and I think that may have been to do with a hormone?' 'What about growing as a child?' queried Julie. 'Don't hormones have something to do with your height? After all you start growing in a big way when you hit puberty.' These recollections of childhood and adolescence suddenly jolted Patrick's memory; he had had a friend as a child who was diabetic. The boy had had to go home and inject himself every day with insulin, in order to keep the sugar levels in his blood on an even keel. 'Isn't insulin a hormone as well?' he queried.

A picture began to emerge, from people's anecdotal knowledge, of a huge variety of hormone-related processes: wellbeing, growth, reproduction, vigour. Hormones appear not only to be involved in physical actions, such as taking flight or producing milk, but also seem to affect our emotions, evoking feelings of fear or happiness or sadness. They clearly move around, affecting all parts of our bodies: our skin, our reproductive organs, our brains. They also appear to have some sense of timing, coming and going with the months and years as we emerge into life, grow to adulthood, reproduce and grow old. With this vague sense of their importance and a much stronger awareness of how little they knew, the group decided to seek out an expert who might explain the basics and address some of their questions.

By good fortune an endocrinologist at a local hospital, Dr Maralyn Druce, was able to find a slot in her busy schedule to meet the group. One evening, in the private clinical room of a white-coated but friendly expert, the group began to discover what an endocrinologist does in her daily work. The first of many revelations was that Maralyn combined the full authority of a scientist with the more personal, empathic approach of a health practitioner. It turned out that her weekly routine involves the same duality. Some days she would be running a regular clinic, treating individual patients with all their complex needs; on other days she would pursue her research on appetite hormones in the laboratory. She provided an inspiring example of how health professionals can combine scientific development with direct responses to human need.

The most fundamental question was answered straightforwardly enough, but nevertheless came as something of a surprise. A hormone is simply any substance produced in one part of the body that travels, usually via the bloodstream, to have its effect somewhere else. It is not defined by its effects, whether these be altering moods or controlling temperature. Nor is it a class of substances, such as proteins or sugars, nor linked to a particular part of the body. It's simply defined by its function as a messenger. As a consequence of this rather broad kind of definition, various ways have developed for classifying the various types. One way is by the family of substances they belong to, for example steroids or proteins. Another way is by the kind of structure they interact with in the body (known as a receptor).

The word 'steroid' acted as a point of reference for people in the group. It was a familiar word, but, as Sonya put it in a debriefing session after the visit, 'What actually is a steroid? I know that when my doctor prescribes a steroid cream for me she always encourages me to spread it thinly and stop using it as soon as possible. It sounds like a dodgy substance!' 'What about the doping issue, too?' inquired Julie. 'Why do athletes take them? Steroids seem to have a pretty bad press. What exactly are they?' These questions brought to the fore an even more fundamental concept in chemistry – one that enables us to make sense not only of hormones, but of the whole array of chemicals we hear about in everyday life. It turns out that chemicals don't just differ from one another in arbitrary ways; there's not just one long list with indecipherable names. Instead, chemicals come in families and there are, as you might expect, important similarities between members of the same family.

Steroids are an example of a family. The illustration below (Fig. 8.1) shows the structure of the molecules of two different steroids, oestradiol and cortisol. Even without knowing what the various symbols mean, it

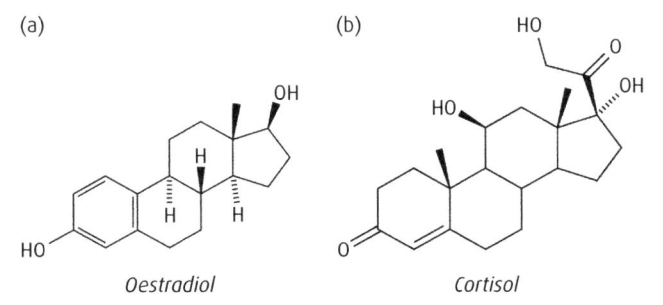

Fig. 8.1 Structures of two different steroid hormone molecules

is clear that these molecules share a common structural element. The hexagon and pentagon shapes represent groups of six and five atoms of carbon. This flat, tessellated appearance of the molecules is characteristic of steroids; the various kinds of steroid share this core structure but, as the diagrams show, differ outside it. Like family members they share some features but differ in others.

Other hormones belong to completely different families, sharing quite different kinds of chemical structure. One important family, the peptide hormones, comprises molecules in the form of long chains, rather like a beaded necklace. Each link in the chain is itself one of a family of chemicals, known as amino acids. Some of these amino acids are well known to careful dieters – they need to be included in what we eat as our bodies cannot manufacture them themselves. In some of these so-called peptide hormones the chain of amino acids is long and wrapped up into a globular shape, rather as the beads in a necklace may curl up in the palm of your hand. A well-known example of this globular type is insulin, the hormone that regulates levels of sugar in the blood. It is malfunctioning of this hormone that lies behind the condition of diabetes. The insulin molecule is much larger than steroid molecules and has a quite different appearance. The size and structure of the various types of hormone differ quite profoundly from one another – their only similarity is in their role as messengers, carrying signals from one part of the body to another.

The molecules of biological chemicals are often quite complex structures – but, as the example of hormones shows, they are usually made up from a limited range of chemical building blocks. These chemical groups, as they are called, give rise to the complicated names you see on pharmaceutical and bathroom products and food labels. The commonly occurring phrase 'methyl', for example, refers to a group comprising one carbon atom and three hydrogen atoms; similarly, 'amino' means a group comprising one nitrogen atom bonded to two hydrogen atoms, while 'carboxyl' refers to one carbon, one hydrogen and two oxygen atoms.

'Well, thanks for explaining what you mean by chemical groups – I suppose it helps to simplify the number of chemical names,' said Mary in a discussion following the group's meeting with the endocrinologist. 'But how does all this chemistry help us get to grips with hormones and how they work?' she continued. 'I get your point that there are these different types of substance, all called hormones just because they have an effect somewhere away from the place they are produced, but what do they actually do? Can we talk about a few examples?'

Some examples

In the discussion at the endocrinology clinic it had become clear that not only was the group right in thinking hormones carry out a very wide range of functions, but also that there were plenty more examples to add to the list. Maralyn Druce was carrying out research on hormones associated with appetite. Apparently the gut makes many hormones. Some regulate the speed at which food passes through the system; others reduce your appetite when your stomach is full. The hormones send signals to an 'appetite integrator' in the brain, which alters our level of desire. In the 1990s a hormone was discovered that had the effect of inhibiting our sensation of hunger. Named leptin from the Greek for 'thin', it is produced not in a gland, but in fat tissue. A few years later a hormone with the opposite effect was discovered. Named ghrelin, its job is to signal hunger to the brain by emerging when the stomach is empty and drying up when it is full. Again this results in signals to the brain, and ghrelin acts on the same area as the leptin hormone. It looks as though the two hormones are among a group of hormones which act to regulate body weight. The system seems to ensure that a balance is struck – not too much, not too little. This is a feature of many of our bodily systems; it ensures we maintain a reasonably steady state internally however the external environment changes. Similar feedback systems regulate, for example, our body temperature, clicking in when it falls either below or above 37°C, and our blood sugar level, ensuring there is enough glucose to supply our energy needs, but not too much.

Maralyn confirmed much of the anecdotal knowledge of the group. Hormones are indeed central to the reproductive system (sex hormones), regulation of sugar levels (insulin) and our growth as children and adolescents (growth hormone). She added other crucial functions they mediate, such as releasing breast milk (prolactin), responding to inflammation (cortisol) and the generation of heat (thyroxine). To date, about 50 different hormones are known in the human body. In discussion with Maralyn, it soon became clear that the mechanism of action of each of these endocrine systems is complicated. Some hormones are released by the actions of other hormones. Some interact directly with nerve cells in the brain; others operate in the pancreas, the stomach, the muscles, the blood and other tissues and organs. Some hormones play a role in several quite distinct systems – a somewhat confusing fact when first encountered. An even greater complexity has been revealed by recent research. It turns out that glands, specialised organs that produce hormones, are not in fact the only source; hormones can also be produced in other organs such as the stomach, skin and fat issue.

Despite this inordinate complexity, however, some key questions stood out for the group. 'Why are teenagers grumpy?' was the first question, succinctly put by Mary, a mother of two. 'It's partly linked to the rapid increase in the level of sex hormones at this stage of life' was Maralyn's response – or, as the BBC science website puts it, 'fluctuations in hormone levels are associated with irritability, recklessness, aggression and depression'. Quick to point to the wider context of adolescent behaviour, our endocrinologist also emphasised other non-hormonal factors, including the psychological issues of emerging identity and relations with parents.

A stream of further questions was inspired by these initial insights. 'Can dreams cause hormones to be released?' 'Can hormones affect thought?' 'Do light levels affect our hormone levels?' The last of these was picked up in a subsequent discussion, when a quick search on the internet showed that light levels do indeed have an impact. A hormone called melatonin is produced in the evening and night time, when blue light levels are low, which helps induce sleep. However in the teenage years its release is delayed, leading to later sleeping and waking times – an issue that adds to our understanding of adolescent behaviour and raises questions about the scheduling of the secondary school day. Recent research in the UK suggests that current school start times often force teenagers to wake up and learn while their body is still prepared for sleep. The body clocks of young teenagers can be running up to two hours later than those of adults.

From this plethora of insights into the diverse functions of hormones one fundamental scientific issue begins to emerge: how do these amazing substances actually produce their profound effects? How do they give rise to such big effects given they are produced in small quantities? How do they manage to be so specific about their targets?

Hitting the spot

One member of the discussion group, Julie, has a wonderful habit of saying out loud what many others are thinking. She did so now. 'OK, Andrew, so we now have some idea about what hormones are and some examples of what they do. But how on earth do they get there and do it? Are hormones somehow soaking into the tissues of the body, taking pot luck on where they end up? Do they all spread around everywhere? Or do they travel with an indicator board up front saying "I'm headed for the pancreas"?' Julie's frank admission emboldened others. 'On top of all that, how do they actually make a difference once they reach their destination?' asked Sarah. 'Where did they come from in the first place – are they made in the body?' inquired Mary.

I expect most of us have wondered at some point what happens when we swallow a couple of aspirin pills. They work their way down to the stomach, where they are presumably absorbed – but into what? And how? An even greater mystery is how they end up in the right place. Does the aspirin somehow know where the pain is and head straight for it? To answer these questions the group decided to pay another visit, this time to a university laboratory dedicated to researching cells – the subject of cytology. More specifically the lab specialised in the way in which cells communicate with one another: the topic of cell signalling.

If you remember anything from this part of biology lessons, it may be the idea that cells are essentially similar to one another. In fact the very word 'cell' was coined by the seventeenth-century scientist Robert Hooke because he saw how the microscopic structure of cork resembled the cells of a honeycomb.

The various types of animal cell do indeed have important similarities – they have surrounding walls ('membranes') enclosing a fluid interior and most have a number of minute structures inside them, such as a nucleus and mitochondria. What struck people in the group more strongly, however, were their obvious differences: in particular the huge variety and specialisation of cells in the human body. There are muscle cells, brain cells, heart cells, blood cells, nerve cells and hundreds more. What seems important in grappling with the questions about hormones is how these cells communicate with each other. If one type of cell is manufacturing a hormone in one place and another is responding to it elsewhere, how are they linked?

The group's visit to the cell signalling laboratory revealed that hormones start from the place where they are made, often a gland, and are secreted into the bloodstream, from where they are distributed, quite indiscriminately, throughout the body. They will pass by all sorts of cells during this journey, but for the vast majority of them nothing much will happen. However, in certain specific cases the outside surface of a particular type of cell will contain something that causes the hormone to 'recognise' the cell. This 'something' is a molecule, usually a large one, which juts out from the surface of the cell. These large molecules are called 'receptors' because their role is to recognise and receive smaller molecules as they drift past.

Receptors are, in many cases, protein molecules embedded in the flexible layers of the cell's outer membrane. A protein is a large molecule, usually of a roughly globular shape. The precise shape of each type of protein is unique, with a characteristic landscape of ridges and valleys, clefts and protuberances on its surface. In the right circumstances a particular type of hormone passing by a particular type of cell can latch on to a particular protein in the membrane. If the relatively small hormone molecule exactly matches a surface feature on the large protein molecule, you may have a kind of key that fits a lock. It is this so-called 'lock and key' mechanism that enables just one specific type of hormone to interact with one specific type of cell. That's how oestrogen manages to affect the reproductive cycle but not the appetite, for example.

So now we have an explanation of how hormones get around the body and manage to 'find' the part of the body they need to influence – the brain, the stomach, the ovaries or whatever. They move around the body in the fluids, but have no effect until they encounter exactly the right receptor molecule. They then interact with this and produce their effect. No sooner had this issue been resolved in the lab discussion than the next question arose. 'OK, so now we know how the hormone singles out the right cell to interact with, but once it gets there how do the molecules know what to do when they interact?', as Sonya put it.

In fact the receptors that hormones lock on to don't only sit on the outside surface of a cell; they may also run right through the membrane of the cell and jut out on the inside of the cell as well, rather like a tie running through the wall of a house. When a hormone locks on to a receptor, the two of them alter shape very slightly. This alteration can be enough to trigger off a sequence of chemical events inside the cell that leads to the overall effect of the hormone.

The example of adrenaline

Sonya had heard about the role of adrenaline as part of her therapy training. As she understood it, when a person perceives an external threat or risk certain glands are instructed to produce adrenaline. 'How does this work?' she queried. 'What happens when the threat has passed – where do the adrenalin molecules go?' This useful example is typical of the way in which many hormonal systems work. Adrenaline is created by the body within a particular type of cell found in glands just above the kidneys: the adrenal glands. These cells contain little sacs which contain the adrenaline molecules. The adrenaline is actually manufactured inside the cells by a chemical process involving various enzymes.

Once the adrenaline has been manufactured, it is released into the bloodstream. From here it is capable of acting on almost any part of the body, but the effect it has will vary according to the particular tissue involved. For example it relaxes muscles in the airways of the lungs, but tenses up muscles in tiny blood vessels. The hormone is well known for the extra power it lends the body in times of fear or stress by increasing blood flow and muscle tension; these in turn help a person to flee more rapidly or to stand and fight. It is also linked to arousal of fear and other negative feelings.

As Sonya had realised, however, an important question remains: where does adrenaline go after the threat has passed? It would be pretty uncomfortable if we had to live indefinitely with a sense of fear and readiness to fight long after the need had passed. How would we ever return to more normal, relaxed feelings? The answer to this brings to the surface another important underlying idea about the chemistry of the human body (and of other living things). Chemical reactions are occurring on a massive scale all the time: we live in a constant ferment of building up, breaking down and dispersal of molecules. In the case of adrenaline, for instance, the molecules are removed from the system once they have done their job. They are either reabsorbed from the bloodstream back into cells or simply broken down into their component parts by enzymes, to be re-used elsewhere or excreted.

How does all this help?

The story of hormones is long and complicated, but also endlessly fascinating, relating as it does to so many aspects of our lives. Numerous individual hormone molecules, each with its own set of interactions; countless cells distributed throughout the body, each with its specialised function; multiple receptors waiting to be activated by their special hormone: there's plenty of detail to contend with just to understand how they work. Discovering how they affect us opens up a host of other stories: growth, reproduction, appetite, fear and happiness, to name just the most obvious areas of influence. But, in the face of so much detail, what can non-specialist people usefully take from all that we have learned so far?

Just exploring a few examples and looking at the mechanisms involved reveals the intricate drama that's taking place throughout our bodies, every second of the day, and through the night too. Special molecules are being manufactured in special places, rising and falling according to demand. Messages are circulating throughout all parts of the body via the network of blood vessels. Levels of vital substances – sugars, proteins, neurotransmitters – are being monitored and constantly re-balanced to keep us on the straight and narrow. Most extraordinary of all, the key changes required for each stage of our lives are gradually being introduced and phased out as we grow older. This realisation heightens one's sense of awe and respect for the working of the body, even in its most steady and apparently uneventful periods.

Being aware of hormone action has further consequences for how we see the social world around us. Human behaviour is, of course, subject to a host of different influences: social, cultural, political and economic, to name but a few. But insight into the role of hormones at different stages and moments in life adds to our understanding of ourselves and how we behave. People on the verge of hormonal change are likely to behave in unfamiliar ways. Riskiness in teenage behaviour, irritability when hungry, rashness in making decisions: all are to some degree subject to the influence of hormones.

Intrigued by their new insights into the world of hormones, the discussion group wanted to learn more about their role in medicines and treatments. Sonya, with her professional interest in body therapy talked about her understanding of oxytocin. Apparently stroking the body and hugging tend to stimulate production of this hormone. Sonya's reference to oxytocin, a hormone whose production is stimulated by stroking the body and hugging, is an interesting example, as research indeed shows an association between oxytocin levels and feelings of wellbeing. Tongue-in-cheek, Julie immediately responded 'Why can't you buy oxytocin in the supermarket?' – a question that raises important points about drug development.

Many hormones are now quite well understood as chemicals, even if their precise mode of operation is not always fully understood. The dopamine molecule, for example, is well characterised; people suffering from Parkinson's disease are known to have lower levels of it due to the death of the cells that make it. Drugs based on dopamine are used to treat the condition, but they don't have the same effects for all patients and there are side effects for some. The hormone serotonin is also well characterised and its association with depression well documented; but treating depression with drugs that increase serotonin levels is a matter of controversy among scientists.

As we have seen, hormone action is in fact a double act. The receptor molecule to which the hormone attaches is as important as the hormone itself. The receptor sits on the surface of the particular type of cell where the hormone action is needed – the muscle cell, liver cell or neuron, for example. The perfect fit of hormone and receptor molecules ensures the action is specific – just this cell, just that hormone. The lock-and-key mechanism depends crucially on both having the right key (the hormone) and the right lock (the receptor).

As you might imagine, a system of such intricacy is quite capable of malfunctioning, and if it does ill health can be the result. And there are many different ways in which it can go wrong. In one type of diabetes, for example, the pancreas fails to produce sufficient insulin molecules. As a consequence blood sugar levels are not properly regulated, causing excess glucose in the blood. In other conditions the receptor molecule itself may not be functioning properly. Receptors are usually large protein molecules. These are made by the body, based on information coded in the genes. It's easy to see that it only takes a simple defect (or mutation) in a gene to give rise to a malfunctioning receptor molecule, which is then unable to respond properly to its partner hormone. It now looks as though this kind of fault may be at the root of some hereditary disorders, including autism. 'Could we not treat receptors if they are not working properly, rather than giving more hormones?' asked Sonya perceptively. But reaching the millions of receptor cells and somehow modifying or replacing them all seems a bit of a tall order at the current stage of science. Yet who knows where future research on replacing defective genes – the blueprints for making proteins – will ultimately lead?

Another way in which drugs can be designed to interact helpfully with receptors is not by enhancing their action, but conversely by blocking them. Indeed this explains the name of the best-known class of drugs for heart conditions: the beta blockers. These molecules have the special ability to attach to the receptor for adrenaline in the heart (and other tissues) and to remain there, thus blocking the site for adrenaline molecules. As a result the effects of adrenaline on the heart – the 'fight or flight' response – are reduced, easing the heart.

Whether designed to enhance or inhibit hormone activity, the task of developing a new drug is almost always a long haul. The journey from scientific discovery to approved medication is long and frustrating, maybe taking decades. The known structure of a hormone molecule may be the inspiration for creating new molecules that either mimic its action or, conversely, block it. Any putative drug has to survive many challenges before it is ready to be used in treatment, however. Is it toxic? Will it survive in the stomach? Will it last long enough in the bloodstream? Is it effective for the target cells? Does it have unwanted effects on other cells? Carefully controlled trials are required to test each of these challenges. These may raise false hopes or lead to dead ends which, even when interim results look positive, may mean further delay before new drugs are routinely available. It's for this reason the pressure is now on to develop new kinds of antibiotics before the evolving resistance of bacteria renders current ones completely obsolete. Powerful new methods for improving the chances of finding effective medicines are being developed today, based on our quite recent understanding of the human genome. These raise hopes of new drugs and treatments tailored more closely to the individual.

Conclusion

The searing effect of hot flushes launched this eventful excursion into the world of hormones. It's this very immediate and personal effect of hormones on our lives that makes the topic particularly compelling. With so many different kinds of hormone active in our bodies, causing such a range of effects, it's impossible to summarise them in one short paragraph. If you have an interest in any particular one it's worth browsing the internet or reading up in the library to find out more about it.

What this chapter has shown is that underlying the great diversity of bodily effects, the basic mode of action of different hormones is similar. Their production is triggered off in some specialised cell, in response to an external stimulus. They get released, usually into the bloodstream, and circulate. At specific places they recognise some receptor molecule and lock on to it. This docking action is the start point for the sequence that leads to the familiar responses of our bodies: releasing eggs, building bones, tensing muscle or dilating arteries. Underlying our daily activities is a truly remarkable system, quietly and subtly shaping our lives.

9
Reflections on Molecules and the Body

'It's just unbelievable!' exclaimed Rosie, giving vent to the unexpressed thoughts of the whole group. Discussion had turned to the precise way in which hormone molecules circulate round through the bloodstream, then dock into a specific cleavage on particular protein molecules on target cells in the body to exert their effects. 'It's impossible to imagine that all this activity is going on all the time just to keep you going every minute of every day – let alone when you're running a marathon or giving birth.'

The concept with which she was grappling was the extraordinary scale and pace of the body's activity at the molecular level. It was a highly pertinent and courageous effort to confront a reality that's almost impossible to visualise. Evidence stacking up over the decades, from countless experiments in laboratories across the world, has given us a strong factual account of how molecules interact in the body. Their sizes are known to be minute, their shapes to be unique and precise and their interactions highly specific and rapid. The sense of intricacy, of perfect, interlocking mechanism, is overwhelming. That interactions at this level are happening in every cell of the body every second – extending muscles, transmitting nerve impulses, fighting off viruses – seems utterly at odds with our usual sense of our bodies as ticking over quietly, stirring from time to time for some special effort.

The challenge of visualising this enormous complexity is ignored in typical discussions about our bodies. 'Ventolin keeps my asthma under control'... 'I get ratty when my blood sugar level's too low'... 'I always feel great after meditating'. We do engage with the chemistry of our bodies to some extent in conversation, but rarely at a microscopic level of detail. Even when we are explicitly learning or teaching about the body at this level of detail, the scale and complexity are easily glossed over. By explaining that 'a nerve cell fires and transmits its signal to the next cell via the synapse', a textbook might be describing a microscopic mechanism clearly enough, yet without conveying its connection to the everyday sensations experienced in the whole body. A pain in the foot or the smell of a curry, for example, will involve thousands of nerve cells firing away in the foot or nose. These then connect up via thousands of other nerve cells to the brain, where thousands of others will be activated in developing sensation and memory. Each of these actions will be completed in a few thousandths of a second, revealing the unrivalled complexity in the body's everyday chemical activity.

But it's not only complexity that raises eyebrows in discussions about molecules in the body. It's also the mechanistic impression you're left with. Every time a protein is manufactured in the nucleus of a cell, a molecule of the DNA double helix is unwound and an RNA molecule is transcribed from it. This then leaves the cell's nucleus and finds a giant molecular complex (ribosome), where it interacts with a multitude of smaller molecules (amino acids) to produce a long chain protein.

This is certainly an intricate process, but not one to be compared to the mechanistic order of an assembly line in a manufacturing plant. In a biological setting huge numbers of molecules are moving around in the fluid medium of cells, driven by diffusion or minute electrical attractions, encountering one another more by chance than design. There's no Fat – or Thin – Controller, directing the affairs of the body's metabolism. It's a statistical consequence of the vast numbers of molecules with varying probabilities of interacting that gives rise to our global impression of an ordered mechanism.

There is more to this concern about mechanism than just its nature, however. When a rush of adrenalin excites the heart or the sound of a love song stirs our emotions, we feel there must be more to this than sheer mechanics! For many people this misgiving makes scientific explanation seem unattractive or even unacceptable. For them, to explain human experience through mechanisms and chains of causation is to diminish it. The idea of chemicals as the active agents in our vibrant, living bodies may seem hard to accept. The very word 'chemical' is so often associated negatively with industrial examples in products such as detergents, plastics and additives. Substances occurring in nature that may be identical are often perceived in a quite different way – as natural, organic or herbal for example, and are rarely viewed as chemicals.

Questions about this contrast between naturally occurring and artificially produced chemicals often arise in discussion – for example,

'Is the iron in our blood the same as in a railway track?' Indeed it is: chemicals are the same however they are made (provided they are free of impurities). Caffeine, for instance, is the same molecule, whether synthesised or extracted from plants. Learning this can be something of a surprise. Even more of a shocker is to discover (as scientists did in the last century) that not only are molecules identically the same whether they occur naturally in our bodies or are made artificially, but that *all* the atoms that make them up were originally created in exactly the same way. Furthermore they were not even created here on Earth, but in the interior of distant stars that have long since exploded. Truly all is made of stardust!

My own feeling is that different perceptions of how our bodies function need to sit alongside each other. Far from diminishing our appreciation of human experience, growing insight into the molecular processes of the body should enrich us. Discovering and learning about these processes are themselves creative and aesthetic experiences, which should be available to as many people as possible. To understand the molecular basis for visual perception or athletic performance does not detract from the feelings we have for these activities, nor does it compete with other cultural explanations. However clear we may become about the molecular processes linking the retina and the brain, the responses we have to what we see and imagine are as individual and ineffable as our very sense of self. Studies of culture and aesthetics, society and history work synergistically with scientific studies to enrich our understanding of ourselves and our communities. Indeed many cultural and artistic endeavours today draw heavily on scientific developments, whether in identifying the DNA of ancient peoples, recording the movements of dancers or analysing literary texts.

With these reflections on the extraordinary nature of the microscopic mechanisms in our bodies we turn to another topic in which a question from everyday life led into the fundamentals of molecular biology. In the next chapter we see how a visit to the doctor's surgery triggered off discussion about viruses and bacteria.

10
Bacteria, Viruses and Antibiotics

'What's the point of viruses?' interjected Sally with an edge of exasperation in her voice. These five simple words captured what everyone in the group was thinking. They had just learned that all these pernicious microbes ever do in life is break into some innocent cell, grab hold of its reproductive machinery, commandeer it to replicate themselves and then push off, obliterating their host in the process. This insight into the reproductive cycle of viruses emerged in the midst of a discussion about the causes of disease.

It had all started after Malcolm and Lucy had been to a talk by two scientists, not on viruses, but on the related subject of bacteria and tuberculosis. There have recently been some worrying statistics showing that the disease is actually on the increase in some of the poorer parts of the UK. Growing resistance to antibiotics was becoming a major concern for TB and for bacterial diseases generally. The discovery of the antibiotic streptomycin towards the end of the Second World War had marked a huge step forward in the treatment of TB. It was the first effective drug treatment for the often lethal disease, which affects the airways of the lungs. Previous treatments had relied on exposing patients to as much fresh air as possible – hence the sanatoria in Switzerland for those who could afford it.

Shortly after the First World War, in 1921, the first TB vaccine (known as BCG) had been introduced, and had had a major impact on the incidence of the disease. Twenty-five years later treatment of TB patients with the antibiotic streptomycin had had a similarly powerful effect on the mortality rate. Since the 1980s, however, just as hopes of eliminating the disease completely were rising, it has become apparent that strains of the bacteria are evolving that are resistant to the antibiotics used to combat it.

Antibiotics

'Before we go any further,' interjected Jean, always keen to sort out the basics before getting deeper into a topic, 'what exactly do we mean by antibiotic?' The question proved popular. 'Yes, surely the terms "anti" and "bio" just suggest these drugs are hostile to life in general,' said Helen, looking carefully at the etymology of the word. In fact, the term had been originally introduced as an adjective in the late nineteenth century to describe threats to microbial life generally; its use as a noun for drugs that worked against microbes dates back only to the 1940s. Interestingly, dictionaries today seem a little unclear about the precise meaning. One describes them as 'drugs that cure illnesses and infections caused by bacteria'; another, more broadly, as 'substances capable of destroying or inhibiting the growth of microorganisms, esp. bacteria'. For practical purposes the word is used today to refer to substances that kill off bacteria in particular.

Meanwhile, back to the talk about TB. The key point impressed on Malcolm had been the new problem arising today: the growth of bacterial strains resistant to antibiotics. The effectiveness of streptomycin has been particularly compromised in this manner. To deal with this a range of new and different antibiotics have been developed in recent decades; some are effective in certain cases, others in different ones. So the common practice today is to treat patients with several drugs simultaneously, typically a cocktail of four drugs.

'That's odd,' commented Sally. 'I had rather thought that a bug could either be killed or it couldn't. How can one antibiotic kill it off, but another one can't? Is one drug entirely different from another or just a variant?' In raising this question Sally had lifted the lid on the deeper issue of what antibiotics actually do to a bacterium.

Bacteria

Microscopic though bacteria may be, they are nonetheless relatively complex life forms. They are single cells with many component parts, each carrying out a different function as with all types of cell (Fig. 10.1). For example, the cells of bacteria have outer walls to insulate themselves from the environment, an interior place called the nucleoid where their genetic information (DNA) is stored and a place where the vital proteins are made (ribosomes). Most intriguingly, the cells of bacteria have

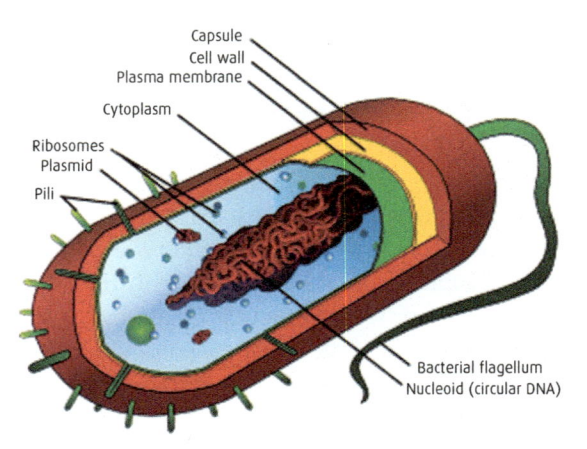

Fig. 10.1 Diagram of a bacterial cell

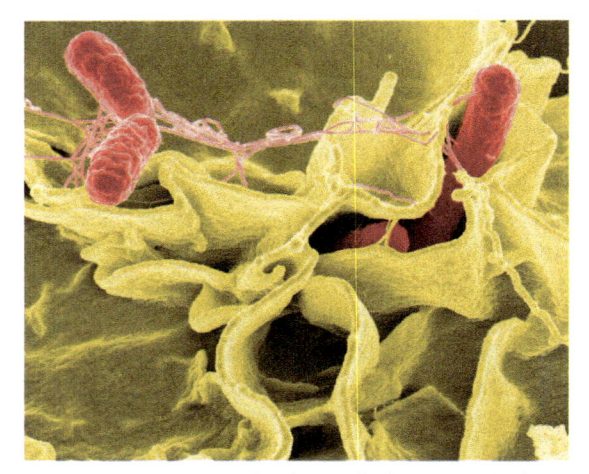

Fig. 10.2 Microscope image of *Salmonella* bacteria (red) among human cells

an external whip-like structure called a flagellum. This enables them to move around actively rather than drift aimlessly in the surrounding currents.

Without wishing to sound too militaristic, it's not hard to see how this relatively sophisticated structure offers several possibilities for attack. Where a bacterium poses a threat to health, the job of an antibiotic is to prevent it from functioning and, above all, from

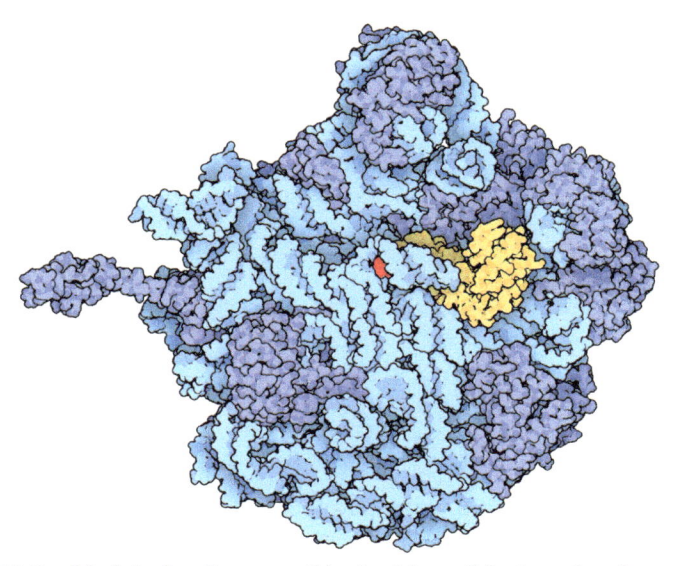

Fig. 10.3 Model of a ribosome (blue) with antibiotic molecule attached (red)

reproducing itself. As it turns out the various antibiotics developed over the decades do indeed interfere with bacteria in a range of different ways. Streptomycin, the first TB antibiotic, acts by inhibiting the production of proteins in the TB bacterium. The drug molecule attaches itself to the place where proteins are manufactured (ribosomes) and stops the creation of vital proteins the bacterium needs to survive (Fig. 10.3).

'What an amazing thing,' observed Rosie. She added, in her characteristic probing way, 'If the antibiotic can do this to a bacterium, why doesn't it do the same in the human cells that are all around?' It turns out that streptomycin is able to do its wonderful work against bacteria without damaging our own cells because the protein-making parts of human cells are fortunately structurally different from those in bacteria, so the drug does not affect them.

Other TB antibiotics, however, work in completely different ways. One, for example, interferes with the production of material to build the cell's membrane, while another inhibits production of the molecules that provide energy for the bacterium. It is remarkable how detailed our modern knowledge is about the ways in which these drugs work. However, it would be misleading to suggest that they have always been designed specifically with one vulnerable spot or another in mind. The possibility of designing drugs in such a precise way has only arisen in recent decades, thanks to the rise of technologies that enable the precise structure and functioning of molecules to be pinpointed. Traditionally the development of antibiotics and other kinds of drug has been much more hit-and-miss. Laborious testing of countless alternative substances was the common method, carried out with the hope of a fortuitous encounter between the test molecule and some target in the bacteria – a process dubbed 'molecular roulette'. The drug itself may have been understood and its effects on tissues observed, but rarely were the two linked by a clear understanding of the causal mechanism.

With this insight into just how specific the interaction of each particular drug is, we can see the advantage of multi-drug treatments. Just because a bacterium may be able to resist an attack on its walls, it's unlikely also to be able to resist interference in its protein machinery – and even less likely to be able to defend its energy-producing function on top of both of these. In other words, the chances of a bacterium being resistant to all three independent assaults are very low. That's what makes the multi-drug cocktail effective against TB.

It was Rosie, fired by her natural inquisitiveness, who put the question on everyone's lips: 'How do these bugs become resistant in the first place? They seem to be engaged in some kind of struggle with the antibiotics.' This well-timed question brought to the fore one of the central challenges in understanding biological systems: the question of purpose or, in technical language, of teleology. It is almost impossible to describe a biological process without ascribing a sense of purpose to it, wittingly or not. By using everyday forms of language we inadvertently imply intentions behind biological actions. I have done so in the paragraphs above – antibiotics 'interfere with' bacteria and bacteria respond by 'resisting attack'. Each of these verbs suggests conscious intention – yet antibiotics are just molecules doing what molecules do, while bacteria are simple cells with no capacity to think or plan. So apologies for the unfortunate impressions that this kind of language gives; but let's also defend its use in cutting down on long, dull sentences that lose the reader in their pursuit of precision.

Certainly bacteria and antibiotics have no intentions – no brains with which to wish destruction on an adversary. They simply are. The antibiotic is a molecule, a type of chemical, not a living thing; it remains as it is throughout a biological process, as long as it isn't broken down. A bacterium, on the other hand, is a living entity. It consumes, it creates, it reproduces – and in this latter capacity we've hit upon its long-term means of survival. The bacterium produces offspring, so inevitably it evolves. It doesn't intend to, it isn't looking for a better life, seeking to defend itself against rogue antibiotics. Evolution just happens, as it does with all living things. As one generation gives way to the next, minute and infrequent changes occur in the genetic material, often at random. Each generation inherits some changes, most of which make no overall difference to the offspring. A few, however, may confer some advantage or disadvantage. If those individuals advantaged in this way prove better able to survive and reproduce themselves, then more of them will be present in the new generation. In this way, over many generations small beneficial changes become embedded in the population – for bacteria as for any other life form.

For us humans, changes of this kind have occurred extremely gradually, over hundreds of thousands of years. But the lapse of time between each generation of humans is particularly long. Imagine the difference with bacteria, capable of producing a new generation within 10 or 20 minutes, hundreds of thousands of times faster than humans. At this kind of pace bacteria simply evolve new capabilities over periods of time that are very short by human standards. Foremost among their new capabilities are bound to be mutations that help them survive, in particular ones that counter the actions of antibiotic molecules. It comes as little surprise, then, that current antibiotics are increasingly encountering new variants of bacteria, no longer susceptible to their actions. No wonder the hunt is on for new kinds of antibiotic molecule.

This insight into the action of antibiotics made a lot of sense for Helen. She had discovered early in life that her body was allergic to penicillin; she would come out in a rash and feel itchy all over. Despite this, however, she was not allergic to other antibiotics. She now saw that this is just what might be expected, given that each antibiotic operates in a distinct manner. Antibiotic molecules are quite different from one another, both in composition and shape, as the models of two molecules in the illustration show (Fig. 10.4). Here each ball represents an atom, with each colour indicating a particular element.

An allergic response occurs when a particular molecule is recognised by the body's immune system and treated (whether justly or not) as foreign to the individual. The act of recognition occurs when the offending molecule fits into a crevice on a receptor molecule in the body. A crevice that neatly accommodates one of the molecules depicted above would scarcely suit the other. An allergic reaction is a response to a specific molecule; you'd hardly expect a body to react to all antibiotic molecules in a similar way, given their great diversity of forms.

(a)

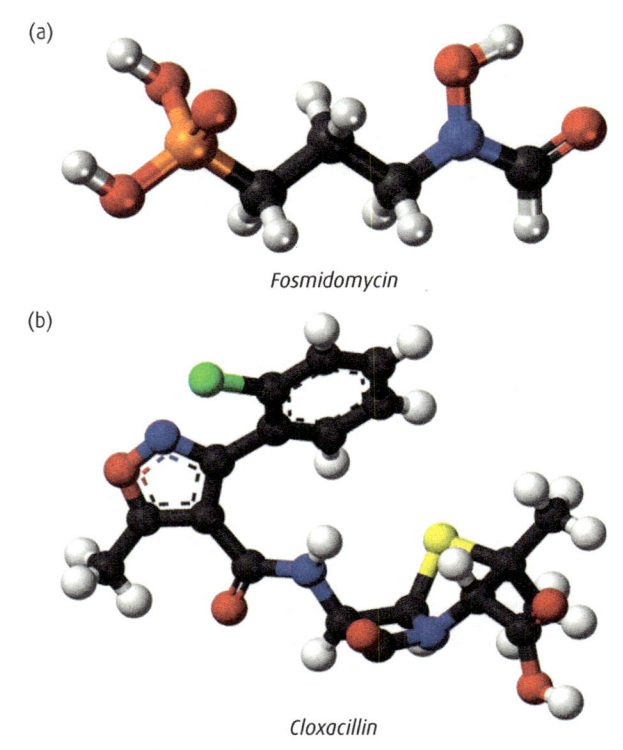

Fosmidomycin

(b)

Cloxacillin

Fig. 10.4 Models of two different antibiotic molecules

Viruses

Talk of antibiotics and allergies reminded Sally of her recent visit to her GP, during the height of the winter coughs and colds. She'd been suffering from an infection that had gone on for over two weeks, finally deciding to take time off work to see her doctor as it didn't appear to be clearing up. The doctor's verdict: it's just a virus, wait and see if it goes away. 'I asked if there was anything I could take, like an antibiotic to speed things up,' she explained, 'but was told that antibiotics don't work for viruses.' Sally had touched a nerve, as almost everyone in the group had experienced this. 'Why don't they work?' was the common plea. Lurking beneath this question was a more fundamental one: what exactly is a virus? Helen had the first shot at answering this. 'Is it just a thing that doesn't respond to antibiotics?' she suggested. 'Maybe it's not a type of organism at all, but is just defined by its behaviour rather than its structure.'

Helen's guess turned out not to be true; viruses are in fact tightly defined entities. The word 'entity' is chosen deliberately here because, unlike bacteria, viruses can't be described as organisms; in fact they cannot even be said to be alive. They don't have the means to reproduce themselves autonomously. What they do have, the key to their potentially devastating impact on us, is the means to hijack the reproductive machinery of others. To be precise, they are able to enter living cells and commandeer the cells' capacity to reproduce. Viruses are structured in a particularly simple way; they consist of a long, thread-like molecule of genetic material covered with a coat of proteins – the absolute minimal structure. The genetic material is in some cases DNA, but more usually is the related molecule, RNA. This contains all the information needed for a virus to reproduce itself. A model of one type of virus is shown below (Fig. 10.5). Called tobacco mosaic virus, it infects tobacco plants. Its simple, rod-like structure shows how protein and nucleic acid (RNA) molecules can combine to form a larger virus structure.

Other types of virus have a different, often more complicated, architecture, but share the same basic combination of genetic material (DNA or RNA) and proteins.

To reproduce, the virus has to attach itself to a cell in the body of its host, then pass its genetic material, a long, thin molecule of RNA or DNA, into the cell. From there, the RNA/DNA can cross into the nucleus of the cell, where it enters the normal replication machinery of the cell. The cell makes no distinction between material from the external virus and its own internal DNA. Unwittingly it reproduces the virus's genetic code. As this alien genetic material passes back into the cell it once again enters the cell's machinery, this time to produce the proteins needed to

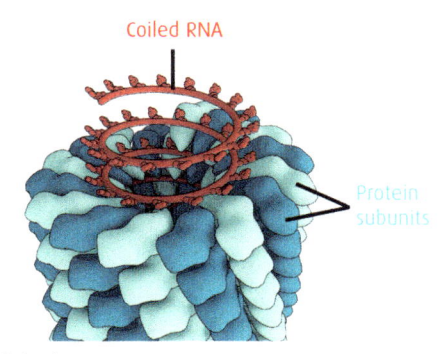

Fig. 10.5 Model of a virus

make a new virus. Protein and RNA, freshly minted thanks to the apparatus of the host cell, then assemble themselves in an orderly fashion to form an entirely new virus – an exact replica of its 'parent'. With its own capacity plundered by the invading virus, the cell is destroyed; its outer membrane breaks up, and a new generation of viruses breaks out into the surrounding medium. It only takes another cell to pass by – for a freshly minted virus to attach to – and the whole process is repeated. In this way viruses are able to multiply at great speed, expanding their own population while simultaneously destroying the population of cells upon which they had relied. The description of this apparently wanton and purposeless life cycle drew gasps of dismay from the group. 'Why doesn't the cell defend itself against attack by viruses?' asked Rosie, expressing a kind of microbial solidarity. 'Doesn't the immune system help?' added Sally. 'Can't it mount an attack on the virus or could it help the cell instead?' 'Why doesn't the cell recognise the DNA of the virus as different from its own?' asked Jean, keen to support the anti-viral sentiment.

It was Rosie who chose to follow on from the story of the virus's reproductive cycle by wondering about the next stage. 'So the virus kills the host, then what?' she asked. Jean's response came swiftly by reflecting on an everyday point: 'They spread, you know: sneezing, coughing and so on...a contagion.' This idea about the spreading of disease reminded Malcolm of something he had read. 'Apparently you get different types of disease in parts of Africa where the populations are very scattered compared to Europe. In thinly populated places, you tend to get water-borne or insect-borne diseases because viruses need the close living of urban areas to get around through sneezes and physical contact. I think that's a reason why TB didn't get a hold in the early USA,' he recalled.

'So viruses just get inside cells, replicate themselves at their hosts' expense, burst out, get sneezed on to another cell and do the same again, and again, and again. What's the point of them?' Sally expressed her reactions with distinct exasperation. She expressed once again the difficulty we tend to have in fully understanding meaning and purpose in biological processes. It's not that viruses really 'attack' or 'invade'; they don't seek to destroy cells. It's just that those versions that do so are more likely to replicate successfully – and as a result are going to be found in greater numbers in successive generations. In this statistical way, they come to be the most commonly occurring, for better or worse. In reaction to this rather bleak portrayal of viruses, Helen tried looking for a more positive angle. 'Is there such a thing as a "good virus" equivalent to the so-called "good bacteria" we hear about?' she asked. By chance Jean knew something about this. She had been listening to a radio programme about new developments in which viruses are being engineered to deliver medical treatments of one sort or another.

Viruses are indeed proving useful in several ways, thanks to their special ability to make their way into the interior of normal animal or plant cells. An example of this type of treatment is gene therapy. Sometimes a gene is faulty, and this fault can be passed on from one generation to the next. The main job of a gene is to act as code for the production of the proteins upon which every living thing relies. Proteins both provide structure for living tissues and drive chemical processes. If a gene doesn't carry precisely the right code, it may not be able to produce proteins normally. Gene therapy works by inserting a normal copy of the human gene in place of the faulty one. To achieve this, researchers have had to find ways to get the new healthy gene to the target cell without arousing the body's defences and, once there, to penetrate its outer membrane to replace the defective gene.

Viruses are particularly good at both of these tasks. After all, that is why they cause disease: they are able to both elude defence mechanisms and burrow into cells. That's the way they have evolved. To take advantage of this ability, recent research has focused on modifying viruses to eliminate their usual detrimental effects. Where this has proved successful the modified virus has been able to pass through the body and enter the target cells without destroying them. A copy of the normal human gene is inserted into the DNA of the virus, which goes on to introduce it into the target cells. These cells are then able to use the new healthy gene to produce the proteins the body requires. Exciting though this new form of therapy is, many challenges lie ahead in making it safe, effective and readily available.

Viruses are used in other beneficial ways too. For example, by modifying them to carry a fluorescent dye they are able to show up precisely where cancer cells are located – a huge aid to radiography. In pharmacology, research is being carried out to see if modified viruses can be developed to deliver drugs directly to the target cells that need them, rather than distributing the drug everywhere in the body. In all these situations it is the specific ability of a virus to evade the body's defence mechanisms that makes them so promising as agents for health-giving treatments. So yes, viruses can indeed be adapted for beneficial ends, whether or not we call them 'good viruses'.

Anti-viral drugs

Frustrating as Sally might have found her doctor's refusal to prescribe an antibiotic, the reason behind the decision is understandable. Destroying the living cell of a bacterium is an utterly different mission from preventing a virus from reproducing itself. A bacterium and a virus are quite different entities. Using an antibiotic to stop a virus would be as inappropriate as using scissors to build a brick wall. Helen saw this point as something of a challenge: 'We've seen how antibiotics interfere with vital processes in bacteria, so couldn't some other types of drug do the same for viruses?' It was a good point. As already explained, antibiotics work in many ways. Some break up bacterial walls; others prevent them manufacturing the proteins or producing the energy they need. There must be some way in which a drug could intervene in a virus's reproductive process to stop it multiplying.

It turns out that developing anti-viral drugs is intrinsically more difficult than developing anti-bacterial ones. This is because viruses don't work autonomously, but exploit the body's own cells instead; if we interfere with a virus's means of reproduction, we risk damaging the host cell at the same time. In other words, our bodies would suffer as much as the virus – something we would experience as a 'side effect'. Nevertheless, in the 1980s research on anti-viral drugs really took off as detailed knowledge of the precise structure of viruses grew, thanks to developments in basic molecular biology. Urgency was added to the task by the parallel emergence of the potentially deadly virus, HIV.

Various molecules have been developed to block the progress of the virus at different stages in its cycle. Some prevent a virus attaching to the host cell, others block its entry through the outer membrane of the host cell. A different type altogether gets into the replication machinery of the cell and jams the production of the virus's DNA. The over-the-counter drug Acyclovir, used for cold sores caused by the herpes virus, is an example of this. Helen is right in that anti-virus drugs are indeed coming on-stream to complement the antibiotics used against bacteria. But, as anyone who's suffered a cold knows only too well, not all viruses can yet be blocked; great advances have been made against some viruses, but there's plenty of scope for further research and development. Progress in this most applied of scientific areas is intimately linked to developments in our fundamental understanding of molecular biology. It's a prime example of how applied and basic sciences are both vital in their complementary ways – the former tackling real-life problems directly, the latter disinterestedly following the path of inquiry, wherever it may lead.

Conclusion

In this chapter we have seen how bacteria and viruses are in fact quite different kinds of thing, although their effects may be equally devastating. A bacterium is an organism, alive in every sense of the word; a virus, not alive in the conventional sense, relies on others to reproduce itself. Typically viruses are much smaller, on average a hundred times smaller than bacteria; they are not visible through an optical microscope, as most bacteria are. As the story of evolution shows, however, size is not all that matters; viruses are by far the most abundant biological entities on Earth. So next time you take time off work to visit your GP, only to be told to sit it out, you may be better prepared and able to take it philosophically. Knowing that antibiotics cannot possibly combat viruses, but your body's defences eventually can doesn't exactly solve your problem, but it may well make it more bearable.

As the past few chapters demonstrate, the human body is a rich source of questions about science in everyday life. Taken together, they show something of the huge range of topics inspired by our bodies: the organs we rely upon – eye, brain, muscles; the systems that connect them – nerves, blood, hormones; the threats we face – viruses, bacteria and disease. Discussion may begin with an everyday concern with health, diet or child-rearing for example, but it often moves on to an exploration of fundamental concepts in chemistry, physics and biology.

The human body is, of course, not the only starting point for inquiries; nor are biological issues the only area of interest and concern. The physical world – the environment, the Earth, the very cosmos itself – also inspires profound questions. Many discussions have begun with questions about these, and such topics also lead on to insights into deeper scientific concepts. One such discussion began with an innocent query about why icebergs float and ended with a philosophical debate about the meaning of nothingness. It forms the basis of the next chapter.

11
Floating and Density

'Why do icebergs float mostly below the surface?' asked Sarah, inspired by a television programme about the frozen north. 'Surely I'm mostly above it when I float in the sea?' Maybe Sarah's body did lie mostly above the briny, maybe not. Apparently it depends on the individual: with more muscle you float lower in the water, with greater lung capacity you float higher.

Thinking about the iceberg, Sally wondered aloud whether the salt has anything to do with it. 'I suppose it makes a difference whether you're in a swimming pool or the sea,' said Michelle. 'Anyway, is there any salt in the iceberg or is it just in the sea?' she added, after a moment of reflection. 'Yes', agreed Sarah, 'what exactly does salt do to ice? After all, they add it to the roads when it snows, don't they?' With these questions and observations flying around, discussion was launched.

Of course the saltiness of sea water varies around the world. It's well known that the Dead Sea, for example, is particularly salty. The cold seas of the Arctic and Antarctic are also salty, but the icebergs in them aren't in fact frozen lumps that have fused out of the seawater. On the contrary: they are lumps of freshwater that have broken off (or 'calved') from glaciers as they descend into the sea. An iceberg is a lump of freshwater floating on top of the salty sea, and typically only about one-tenth of the iceberg is above the water line.

This image of ice floating intrigued Michelle. 'OK, an iceberg floats; so do ice cubes in a glass of water. So the ice must be lighter. What if you have a bottle of water and you freeze it, would it become heavier or lighter?' 'If it's in a bottle, does its weight depend on being in a container?' wondered Sarah, bringing in a different consideration. 'I can't see that,' responded Mary, using logical argument. 'Surely if water got lighter when we freeze it, where would the weight have gone?' 'If it's filled to the brim it can burst,' commented Julie, recalling a disaster when she had once put a bottle of water in a freezer. As she saw it: 'The ice just takes up more space than the liquid. You see it when you make ice cubes – the cubes stick out higher than the water you put in.' It was left to Michelle to put the evidence together: 'The ice and water must weigh the same, as Mary said, but the volume gets larger. The water stays the same amount, but it must expand in the freezer as the ice forms.'

The meaning of density

This dialogue finally clarified for the group some basic aspects of physics they'd never been really clear about. When water freezes into ice, the *weight* stays the same. No more material is added and none is taken away; or, as Mary put it succinctly, 'There's the same number of molecules whether it's water or ice.' But the ice occupies a bit more space or *volume*; as Sonya appreciated: 'The molecules spread out further.' This is where the concept of *density* proves so helpful. If the same amount of stuff takes up more space, it means it must become less dense. This is the reason why ice floats on water: it's less dense. It's just like a block of foam or a piece of wood – it is to do with the stuff that it's made of.

Feeling unclear about the concept of density is quite understandable; confusion is unfortunately built into the very way we use language. We commonly say a pebble will sink because it's 'heavier' than water; equally polystyrene floats because it's lighter. To be accurate we should say the stone is *denser* and the polystyrene *less dense* than water. What we are talking about is a property of the material. Stone or steel are dense materials. Polystyrene or wood are the opposite, but unfortunately there isn't an everyday word for 'undense' so it's easy to slip into calling it 'light'. Used correctly, the words 'heavy' or 'light' cannot be used to describe a material such as wood or steel; they are used instead to describe particular *objects*, such as a table or boat.

'So does this mean wood *always* floats?' asked Sonya, adding, 'Is that why boats are made of wood?' Julie picked up on her first point, thinking about the size of different pieces of wood. 'I suppose a log will float down a river just as much as a twig will. It doesn't really matter if it's big or small; it's the fact that it's made of wood that counts.' This simple observation captured the key point: all wood floats because it's less dense than water, and the reason for this lies in the very structure of the substance of wood – its ingredients are not so tightly packed. Think sponge cake versus fruit cake. 'Ah, but what about a water-logged log, though?' challenged Sarah. 'Wouldn't that be heavier than water?

I mean denser than water? Don't logs like this sink to the bottom of rivers?'

It's true, objects are often made of a mixture of substances. A water-logged log is indeed a mixture of wood and water. Even more striking is the case of boats, which aren't always made of wood; they can be made of plastic, fibreglass or steel. 'OK,' interrupted Sarah. 'I can see that ships today are mostly made of steel, but if it's denser than water, how on earth does a ship float?' The answer is that only part of the composition of a ship is steel; some of it is wood, some plastic, but most of its volume is taken up with air, which has a very low density. So averaged out, the density of a whole ship is less than that of water – that's how it floats. You can see that a ship floats lower in the water as it's loaded up – its overall density is gradually increasing as weight is added. My guess is that whether a water-logged log floats will depend on how saturated it is. It appears that ancient timbers at the bottom of the ocean are denser than water, as air trapped in the wood has been gradually replaced by water. It could be worth taking a look next time you're by a river.

'What exactly do we mean by weight, then?' inquired Sonya. 'When I blow up a balloon, is it getting heavier?' 'Surely it must be,' responded Michelle. 'After all you're adding in more air and air weighs a bit, doesn't it?' This helped to clarify the matter. Weight is about individual objects: this empty balloon, that full balloon, this table, that boat. What is therefore the link with density? One kilogram is the weight of a bag of sugar, and we know roughly what size that is. But if we had a kilogram of polystyrene, it would take up a lot more space – think of the Styrofoam we get in packaging. Both would weigh the same (one kilogram), but the polystyrene will take up a lot more space because it's so much less dense.

'Can we say exactly what we mean by volume? Is it really just the space something takes up?' ventured Sarah, anxiously revealing her uncertainty about something she thought she was supposed to have grasped 20 years ago. Indeed, that's just what it is. The whole interlinked discussion was finally pulled together by Mary, who captured the three main concepts succinctly: 'Volume is how much space something takes up; density is how spread out the molecules are; weight is how much all the molecules weigh.' 'So do all icebergs float with the same amount below the surface?' asked Sarah, reverting back to her original question. 'If they are all made of the same stuff they must have the same density, I suppose.' Correct: all objects made of a particular substance will float with the same fraction below the surface – nine-tenths in the case of icebergs. So logs and twigs of all sizes will float with the same proportion below the surface because they are all made of wood (setting aside the small differences in density between hard and soft woods).

A fragment of school physics was vaguely recalled by Mary. 'The Archimedes principle – isn't that something to do with floating?' she asked tentatively, knowing it was something she had never understood despite being able to recall the name. This piece of physics theory helps us to visualise what is actually happening when something floats. If we imagine a ship afloat, the part of the hull that is under water must be sitting in a space previously occupied by water. Yet the ship is floating nicely, neither sinking nor rising. So the ship must weigh exactly the same as the water that was previously there. 'OK, so the submerged part of the ship replaces the water that was once there, I get that, but how does this help?' asked Sarah. The question threw up something of the value of theory. Archimedes had seen that, if a ship floats, there must be something pushing upwards that balances the weight pushing down. He called this an 'upthrust' on the ship, and saw that it must be the same as the weight of the water that was previously there. This fundamental piece of theory explains the mechanics of floating, and is essential when it comes to designing ships and balloons.

The special case of water

Sarah, less concerned with theory, was keen to bring the conversation back to the practicalities of ice. 'We know that icebergs float on the sea and ice cubes do the same in a glass of water. I suppose that's why ice forms first on the top of a pond in winter,' she conjectured, correctly. As temperatures plummet at night in winter, the air in contact with the pond begins to cool the water at the surface first. As it does so, ice begins to form and, as it is less dense than the water from which it came, it remains floating on the top. Had the ice been denser than water, it would have sunk to the bottom as it formed.

'Water is weird,' interjected Mary. 'You'd expect stuff to get denser when it goes from liquid to solid. Things generally contract when you cool them and expand when you heat them up, don't they? But when water cools down into ice you're saying it expands. Why's that?'

It's true, water is weird (or anomalous, to use the scientific term). It seems strange that water, the most ubiquitous liquid on Earth, actually behaves quite untypically. Molecules of H_2O, the particles that make up water and ice, like all molecules, experience a slight attraction to one another when they are close. It's this that actually holds substances together at all, rather than dispersing as vapour into the blue yonder. To be more precise, vapours and gases do dissipate in this manner, but that's because the molecules are so far apart from one another that their mutual attraction tails off. For liquids and solids, however, the molecules lie cheek by jowl, making these forces of attraction strong enough to keep the molecules bound together. That's what holds a spoon or a brick together as solid objects, or a glass of wine as a liquid. In the solid form the molecules of a substance are locked in place next to each other in a regular array. In a liquid they also lie next to each other, but are free to move past each other, giving rise to the effect of fluidity. For most substances, the space occupied by molecules in the liquid state and the solid state is much the same, so the density of the two forms does not differ much.

The case of H_2O molecules is different. In the liquid form (water) they pack closely together as shown in the left-hand model below (Fig. 11.1). However, in the solid form (ice), the molecules are held in a geometrical array that spaces them a little further apart. As the models reveal, the molecules of ice take up a little more space, leaving gaps, so ice is slightly less dense than water. This is the fact that makes it float.

Discussion groups have an unexpected tendency to absorb new concepts like this alarmingly quickly once they have been carefully explained. This is, of course, tremendous for those who are learning, but for the teacher it's exhausting: no sooner is an idea finally bedded down than a fusillade of further questions is launched. Discussion about the intimate structure of ice molecules proved no exception to this rule. No sooner had the spaced-out nature of ice molecules been explained

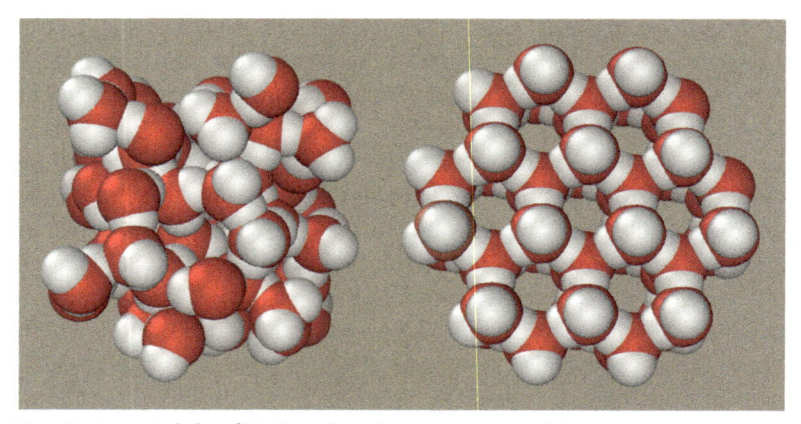

Fig. 11.1 Models of H_2O molecules as water and ice

and understood than Julie demanded in her characteristically inquisitive way: 'Why are the H_2O molecules in ice spaced farther apart than in water?'

The answer lies in the nature of a special kind of bond between two molecules that occurs when a hydrogen atom on one molecule comes close to an oxygen atom on another. This exists because, although an H_2O molecule is electrically neutral, the oxygen part is very slightly negative and hydrogen parts slightly positive. This gives rise to an attraction between neighbouring molecules, bonding them to one another. Such so-called 'hydrogen bonds' are relatively weak, and are additional to the much stronger bonds that hold together the three atoms in an individual H_2O molecule. These weak hydrogen bonds develop everywhere in water, but their fragility means that they are continuously broken in the hurly-burly of the liquid state, in which the molecules are continuously tumbling over one another. As a result the hydrogen bonds have very little effect on the structure in the liquid form (water).

In the ice form, however, where the atoms are no longer rushing around but rooted in fixed positions, the strength of the hydrogen bonds between neighbouring molecules is sufficient to affect the geometry of the ice crystal that forms. These special bonds are slightly longer than the usual distance between molecules, so they have the effect of holding neighbouring molecules slightly further apart than they would otherwise be. This settles the overall structure into the beautiful geometry illustrated earlier, with its wider spacing and open holes. It's this extra space in the ice structure that reduces its density compared to liquid water.

'Wait a minute,' interrupted Mary. 'What are you saying? Open holes in the middle of ice? What's in the holes?' With this unexpected turn of conversation Mary had embarked unwittingly on a new and profound line of discussion: the nature of the void. In between molecules there is indeed emptiness – a vacuum in the heart of solid matter. 'It freaks you out,' she continued, 'that there is just a vacuum between molecules. Are solid objects really full of empty spaces? Still, I suppose if you think of air it's 99 per cent empty space, isn't it?' 'Is it really a vacuum or is it just that we don't know what is there?' she added, somewhat suspiciously. Classical science teaches us to accept the strange truth that there really is nothing in between molecules: the vacuum is all around in space and it's right here, under our noses, in solids and liquids, as well as gases. It gets worse. Even the majority of the volume occupied by an atom is empty. The particles that make up atoms only occupy a tiny fraction of the space – the rest is emptiness.

This is just one of the amazing realisations that creep up on us as we explore science.

How materials can be as hard as iron while the atoms of which they are composed consist almost entirely of empty space is the obvious issue that this revelation raises. Discussion of how materials acquire their qualities is for another day, however; our journey into the anomalous behaviour of ice and water, exploring en route the fundamental concepts of density and weight, is surely enough for one sitting.

Water, in its countless forms, is a boundless source of inspiration for the curious. Icebergs, oceans and ponds inspired this discussion, which has opened up scientific ideas about floating and density and the unusual behaviour of the H_2O molecule. In other discussions people have picked up on the way water moves, the simple things we might notice on a seaside holiday or a leisurely riverside stroll: the ebb and flow of tides, the swirling of river currents or the lapping of waves against the shore. Such an observation provided the trigger for a fascinating discussion one evening, which led the unsuspecting participants from a beach in northwest England to the very forces that bind the universe together. Chapter 12 is based upon this fantastic journey.

12
Tides and Gravity

'Why is the tide so far out at Blackpool?' asked Terry after a trip to the coastal resort one summer. 'I went there on a visit to see the sea, but it was so far out you could barely see it.' Wide beaches are a striking feature of some resorts, frustrating for eager children facing the tedious walk to the water's edge and potentially dangerous too for beachcombers unaware of the speed of incoming tides. 'Is it something to do with evaporation in the summer heat?' Mary queried. 'What about the polar ice cap melting, is it to do with that?' asked Julie, before realising she wasn't sure whether this would cause the sea level to fall or rise. 'That's a point. Does a melting iceberg cause the water level to rise or fall?' wondered Sarah, aloud. 'It's releasing melted water, I suppose, but on the other hand it's vacating all that underwater space it took up.' Joining in the speculation, Rosie offered yet another possibility: 'Maybe the heavy rainfall in the northwest has something to do with it.'

The image of melting ice caps inspired Mary to further thoughts. 'I don't think our Earth is like a glass of water with ice cubes in it, anyway. There isn't some kind of a container. The water just holds on by gravity and spreads out all over the surface, doesn't it? All the water is joined up, so if the level rises it must rise everywhere.'

These initial observations showed something of the complexity of the movements of water. Tides come and go on beaches, rains fall and drain away, waves roll across the ocean and dissipate upon the shore. So explaining the rhythm of the tides, or any other flow, is no simple matter – many factors are at play. A glance at the tide tables in different coastal areas is enough to suggest this. The timing and height of tides seems to vary from place to place, even when they're close together. A narrow entrance to a bay may restrict the flow of water as the tide changes, delaying or even limiting the rise and fall. Malcolm had once heard that 'the main reason tides are so low in the Mediterranean is because so little water is able to flow through the straits of Gibraltar that it never has time to fill up'. The tides can indeed be affected by local geography – narrow straits and inflowing rivers for example. But the general picture, the daily ebb and flow, is, as Mary had suggested, down to the effects of gravity. Furthermore, as she had pointed out, the water across the oceans of the world is indeed linked up in one continuous piece. Mention of the word gravity, common enough in everyday conversation, began to ruffle brows in the discussion group. It was the trigger for a deeper spate of questioning. What exactly is gravity? How does it affect the tides?

Introducing gravity

The importance of gravity is picked up intuitively from an early age, well before the theory is understood. We notice that things fall down, not up. We have weight and leave footprints in the sand where we press down. When we are older we may see images of astronauts walking unusually lightly on the surface of the Moon, and learn that gravity there is much weaker than on Earth. The very word itself imparts a sense of profundity: 'She had a certain gravitas', as we might say of an impressive person. Perhaps those most renowned of all scientists, Isaac Newton and Albert Einstein, are both associated with fundamental theories about the nature of gravity. So it's a serious concept.

Despite its highly abstract nature, gravity is in fact quite a simple concept to grasp for ordinary purposes. It is a force that exists between all objects, due simply to their mass. In other words the amount of matter in objects determines the strength of the force between them. 'Hold on a minute,' interrupted Marian, challenging this apparently simple proposition. 'Every chair and table and human being isn't pulled towards everything else. If there's a force between them it must be pretty feeble.' Indeed, the crucial feature of gravity is just how weak it is; ordinary objects do not pull towards each other as magnets do. The force is very weak indeed, and only really noticeable in the vicinity of massive objects. The pull of ordinary objects, such as tables and chairs, on one another is immeasurably small. But of course the Earth we live on is no ordinary object; it's a good example of a huge mass, so massive that it exerts a noticeable pull on us. This is what we call 'weight': the pull of the Earth acting on the mass of our bodies. It presses down so our feet leave an imprint in the sand.

It presses down on the springs of our weighing scales, turning the dial to reveal our individual weight.

Gravitational force doesn't depend on contact, however: it's not a kind of glue, binding objects together where they touch. It's more like the force of a magnet acting on a paper clip, or a charged-up comb whipping up our hair; it acts over empty space. That's what's so profound about it, the point that Isaac Newton firmly grasped: gravitational force acts everywhere in the universe, every single object pulling on every other one. It's universal, pulling on apples, people, planets, stars and moons – attracting each to each in proportion to their mass. He also realised that gravity isn't equally strong everywhere – it tapers off as we move away from an object, just as the magnetic effect does further away from a magnet.

In the case of our Moon, of course, we've got one pretty massive object fairly close to another pretty massive object, the Earth. So, not surprisingly, quite a strong force of attraction acts between them. Strong enough to distort them slightly, so each is very slightly stretched towards the other; even the rocky Earth and Moon are elongated into something of a rugby ball along the line that connects the two. Mind you, the effect is extremely small, stretching the diameter of the Earth by a mere 30 cm approximately.

However, the effect of the Moon's pull on the sea that surrounds the Earth is more dramatic, as we noted in the case of Blackpool. As Mary pointed out earlier, the oceans resemble a kind of cloak, covering much of the Earth's surface in a continuous thin layer of connected water. Resistant though solid Earth may be to the attraction of the Moon, the oceans are more susceptible. Being free to flow, the water of the oceans heaps itself up when the Moon is overhead, pulled towards it by gravitational attraction.

When Blackpool happens to be directly in line with the Moon the sea is pulled up – a high tide. When it's a quarter turn away from this position, the sea is thinned out, resulting in a low tide. So the level of the tide all depends on where we are on the Earth's surface in relation to the Moon at that moment. But of course the Earth never stands still; it's rotating steadily, presenting new places to the Moon with every passing minute. As the place we're at gradually moves, so the level of the ocean rises and falls. It's a 24-hour rotation but, as the diagram illustrates (Fig. 12.1), the tide is high at any given moment on two opposite sides of the Earth. As Blackpool rotates under the Moon, two high tides and two lows will come and go in the course of a 24-hour cycle.

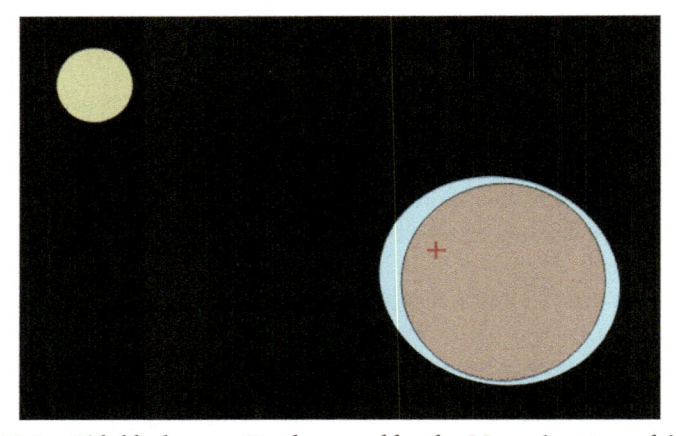

Fig. 12.1 Tidal bulges on Earth caused by the Moon (not to scale)

'I can't see why there should be two tides,' queried Michelle. 'Surely if the Moon's pulling on one side of the Earth, all the water should be dragged that way?' A very fair point and at first sight a baffling one. The explanation comes from understanding what happens whenever any object circles around any other one – a child on a roundabout or a satellite above the Earth. Whenever a thing is orbiting around a point there has to be force from somewhere, always directed towards that point, to prevent the thing flying off at a tangent. We feel it if we grip on to a children's roundabout as it spins round; your arms struggle to hold on and pull you inwards to prevent you being flung off tangentially. Whether for an artificial satellite or a natural one, such as the Moon, the gravitational attraction does this pulling job. Steady rotation occurs when this inward ('centripetal') force exerted is exactly the right strength to maintain the orbit.

Gravitational attraction gets gradually weaker the further away you are. The gravitational pull of the Moon on the Earth is just the right strength to maintain the orbit of the *centre* of the Earth. On the part of the Earth nearest the Moon it's a little bit stronger than needed, however; on the opposite side of the Earth, lying furthest from the Moon the gravitational force is a bit weaker. As a result, where the Moon is nearest the Earth, its excess attraction pulls the ocean towards itself; but where the Earth is farthest away from the Moon, the extra weakness of attraction leaves the ocean bulging out away from the Moon. As the diagram shows, the two bulges mean the tide is high locally on the two opposite sides of the Earth at any given time.

Ways that water moves

Once this rather complex theoretical explanation of gravity and the Moon had been digested, discussion reverted to the everyday reality of the seaside. 'When the tides come in every day, is it always the same water coming back and forth?' asked Terry, capturing a point that many of us have no doubt pondered at some point. 'If it is, how would messages in bottles ever reach the shore?' queried Julie, introducing a new factor into the debate. 'Water is moving all the time; where do tides fit into the wider picture?' asked Mary. 'In fact, what do we actually mean by "water" – is it all the molecules?'

It's true that as soon as we think about the way water actually moves about in the oceans, there are quite a few factors to consider. The bulk of water is indeed a continuous mass of H_2O molecules – plus quite a few others besides, derived from the various minerals dissolved in it as well as any accidental impurities that may have been introduced. The most visible kind of movement is the rising and falling of the waves on the surface. These appear as a kind of swell out in mid-ocean, but closer to the shore they break and crash as their crests overreach themselves. Although it seems to contradict our intuition, waves don't actually carry water from one part of the ocean to another. We see this clearly and unexpectedly when a wave passes over a piece of flotsam: it simply bobs up, moves along a bit, drops down and returns more or less to where it started as the wave passes. The particles of water are rolling round in a large circle, first rising up with the wave, then falling back The appearance of forward movement is due to the profile of the wave shape passing along from each section of water to the next. In the diagram (Fig. 12.2), the peaks (2) and troughs (3) move in the direction of travel (1), but the particles of water simply circle round in deep water (A) and move elliptically in the shallows (B).

Nearer the shore the floor of the ocean begins to drag on the lower part of the water, slowing it down. This causes the upper part of the wave to rush forward relatively, tipping over the crest of the wave and providing great fun for surfers.

So, by and large, waves don't contribute much to the mass movement of sea water. Nor do tides. As the latter ebb and flow, it is indeed largely the same water coming in and out. The effect of the Moon's gravity is to raise the level of the water at a point during one part of the day and gradually lower it the next – rather like a vacuum cleaner lifting the surface of a loose carpet as it moves over it. There is no net flow of water from one region to another. By contrast, wherever there is a bulk flow of

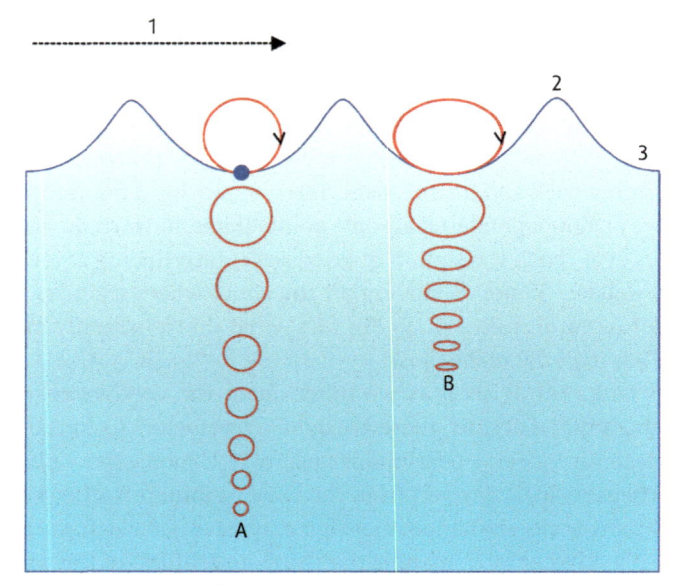

Fig. 12.2 Movement of water in ocean waves

water from one place to another what we notice is a current. Currents can be caused by winds on the surface or by differences in temperature and salinity. Where the water is warm and less salty it is less dense, and so tends to rise. Where it is cooler and more salty, however, its density will cause it to sink. These vertical movements drive large circular currents. Across the globe very large cyclical currents run from warm equatorial regions to polar ones through the depths of the oceans. These exert a large influence on weather patterns by transferring heat energy from the warm equator towards the cooler poles.

The combined effects of currents and tides means the ebb and flow of water in each locality varies. The effects of headlands and bays can make a difference, as can hollows and hillocks on the bed of the ocean. There are also other complicating factors that affect the height and timing of tides, including the tilt of the Earth's axis and the influence of the Sun. The latter has a gravitational effect even though it is so far away, simply because it's so very massive – it's the Sun's gravitational effect that accounts for the more extreme neap and spring tides. Spring tides, confusingly, are not related to the season of the same name. They are simply the more extreme tides that occur when the Moon is either full or the opposite, new. This happens on two occasions in the course of a month when the Sun, the Earth and the Moon are more or less in a line. In this situation the gravitational effects of both the Sun and the Moon add together. As a result the high tides are higher and the low tides lower. Roughly a week later, with the quarter Moon, the Sun and Moon are at right angles to each other, so the tidal peaks, known by the name 'neap', are less extreme.

More on gravity

Tides are clearly fascinating; they intrude on the edges of our land, raise and lower our boats and shape the harbours of our historic coastal towns. Discussion about them often begins with things you observe in everyday life, such as the widening and narrowing of a beach over the course of a day, as Terry had observed at Blackpool. You may even notice how the timing of the peak tide varies from one coastal place to another, or how it fluctuates as the Moon passes through its monthly phases.

In one discussion group, getting to grips with basic ideas about the gravitational effect of the Moon proved just a beginning. Inspired by the explanation, they found themselves asking further, deeper questions. Rather imaginatively, Rosie asked: 'Does the Moon's gravitational attraction lift up sand in the desert like it does water in the ocean?' As she was speaking, Helen began to perceive an even bigger picture: 'Gravity can't just affect sea water surely, it must affect the solid substance of the Earth as well. I can't believe gravity distinguishes between different types of substance – water, sand, rock?' As we saw earlier, this is true – gravity does indeed act on the bulk of the Earth as well as the oceans on its surface, but the bulges and contractions are extremely small, measuring less than one metre. On the whole, therefore, we don't notice the tidal effect of the Earth itself, but presumably the sands of the desert are lifted up slightly. It would be interesting to know how big this effect is.

'What about other planets?' asked Jean, extending the gravitational concept. 'Does the gravity of a planet depend on whether it is made of gas or not?' Sonya took the questioning a dramatic step further, presumably having read or seen something about relativity theory: 'What actually causes gravity? Is it the curvature of space alone?' By probing the ultimate cause of gravity and its contemporary explanation in terms of curved space, the question ushered in a more historical discussion about the two main theories of gravity.

The concepts we use in our everyday lives to understand the way things move around and affect each other date back several hundred years, to the sixteenth and seventeenth centuries. Then the so-called 'laws of motion' were beginning to be worked out by scientific thinkers and experimentalists such as Galileo Galilei in Italy and Isaac Newton in England. They saw that to get something to start moving from rest, or to slow down something that was already moving, some kind of force needs to be acting. We take this for granted when we begin a car journey by pressing the accelerator or stop it by applying the brakes. This concept of force became extended to include less tangible effects, for instance to explain what's happening when a magnet attracts a piece of iron or a stretched bungee holds down our luggage. We associate forces with physical objects, such as engines, elastic threads or magnets, acting on things. For car engines and bungee cords there is physical contact between the forcing thing and the body on which it acts; it's relatively straightfoward to see how one affects the other. With magnets, however, it's not so obvious. What's happening between a magnet and a paper clip when they attract one another from a distance?

This fact – that forces of attraction can operate between things that are not in contact, referred to as 'action-at-a-distance' – is hard to understand and accept. Children are surprised when they first see a magnet attracting a steel toy or a rubbed balloon pulling in scraps of paper or particles of dust. It seems to contradict everyday experience, to be magical. As adults we just get used to these things. On the other hand, a cup falling from our grasp and smashing on the ground doesn't seem surprising, even to an infant: it's something we learn to expect at an early age. Yet with a little reflection even the everyday experience of falling is as mysterious as magnetism. 'It falls because of gravity,' we might say; but this kind of explanation simply defers the question. If you are baffled by this, you are in good company: philosophers over the centuries have struggled with the problem. Isaac Newton, widely seen as the 'discoverer' of gravity, admitted: 'I have not as yet been able to discover the reason for these properties of gravity from phenomena, and I do not feign hypotheses.'

In the absence of a satisfactory philosophical explanation of action-at-a-distance, later natural philosophers took a pragmatic approach. They simply invented a new concept to cope with it: the 'field'. Today this concept is used in practically every branch of science in an everyday kind of way. In effect, it side-steps the philosophical worry by just accepting that action-at-a-distance happens. We routinely invoke the field concept to explain magnetic, electrical and gravitational effects. We say that the presence of a magnet or electric charge results in an area of influence around it, dubbed a 'field'. Within a magnetic field any other magnet or piece of iron will experience a force due to the first one. This is reciprocated, so that each experiences the influence of the other. Similarly an electric field surrounds any electric charge; and other charges in the area will experience a force as a result. So we no longer look for some unspecified medium in the space between magnets or charges; instead we simply accept that influences are felt across intervening space.

'Yes, that's all very well, but we were talking about gravity,' interjected Sonya at this point, faintly impatient with the diversion into magnetism. 'Are you saying gravity's like magnetism?' Mary joined in eagerly with another angle: 'What about a vacuum? Doesn't that make any difference? Isn't there just emptiness between here and the Moon?' The example of magnetism or static electricity is introduced here simply to give a more visible example of forces operating over a distance. For gravity the concept is just the same, but it's a bit more abstract and less easy to visualise. Gravity is also understood today as a 'field', a zone of influence. But in place of the magnet or electrified balloon there is simply a mass – just any old chunk of stuff. Any object at all has mass. It's in the nature of all objects and the particles of which they are made. The more stuff they are made of, the greater their mass.

The big difference is that the gravitational field surrounding any mass is very, very weak compared to the kinds of field we see around magnets and electric charges – billions and billions of times weaker. It's for this reason that the everyday things in life tend to stay roughly where they are rather than rushing off in bursts of mutual attraction. Two people standing one metre apart would exert a gravitational attraction on each other of about 0.00000003500 Newtons – equivalent to one ten-thousandth of the weight of a grain of salt! Gravitational forces that you really notice only occur when immense amounts of matter are involved. That's where the Earth, the Moon and the Sun come into the story.

Massive things such as stars, planets and moons exert a noticeable gravitational pull just because of the amount of stuff of which they are composed. It's the sheer mass of the Earth that pulls things towards it. Cups fall downwards off tables, rivers run downhill and feet are rooted to the floor, all because the Earth is pulling inwards on everything. It's this pull that our weighing scales register as our weight. Weight is simply the force exerted by the mass of our bodies (or anything else) as they press downwards on the Earth. If we were to stand on the Moon, we'd lose a lot of weight in an instant because the Moon is far less massive than the Earth and its gravitational pull is correspondingly weaker – only about one-sixth of the amount. Our weight would reduce by five-sixths (83 per cent) on the Moon scales, but I am afraid that our mass would remain just the same – and with it, our body mass index (BMI).

Moons, planets and Newton

The universe as a whole is composed of massive objects: stars, planets and moons, each exerting forces of attraction on all the others. Given the unimaginable number of massive bodies in the universe, these forces would seem to add up to an impossibly complex set of interactions to try to make sense of. In practice, however, things work out a little more simply. The gravitational field surrounding a mass is not only weak, as we have already seen; it also falls off very rapidly as we move away from the mass. As a consequence, objects are influenced mainly by the masses closest to them. Thus the Earth is mainly influenced by the Sun, very slightly by the Moon and other planets in the solar system and barely at all by more distant objects, such as the stars.

Through these considerations, we have more or less arrived at the great theory of gravitation conceived by Isaac Newton in 1687. He saw that gravitation was universal – the apple that was reputed to have fallen in his Lincolnshire garden was subject to the very same force as that which held the planets in their celestial orbits around the Sun, and the Moon around the Earth. He saw that the strength of this force depended on the masses involved: the gravitational influence around the Sun is bigger than that around the Earth, and this in turn is larger than that around the Moon. At the same time the amount of force also depends on how far apart the bodies are. So the Sun, massive as it is, affects our Earthly tides much less than the Moon. This is because the Sun is about 400 times further away – 150,000,000 km compared to 384,000 for the Moon.

The concept of gravitational attraction goes a long way to explaining the behaviour of the tides (the original starting point for this discussion). Being universal, it also goes a lot further than this. Grasping the two basic ideas in Newton's theory – that there is a force between masses that depends on how massive each is and how far apart they are – gives us insight into one of science's most fundamental ideas. It helps us understand how things move, how the universe is structured, how it began and how it's evolving. For Mary, grappling for the first time with this most fundamental of concepts, a new question immediately arose. 'If all matter is attracted to all other matter, why doesn't it all just collapse into one big blob?' she asked, expressing a difficulty that almost anyone feels who thinks about our universe. What is it that keeps everything revolving?

I suppose the most straightforward answer is that things in the universe have always been spinning. And what is there now in the universe to stop such spinning? Had the universe somehow exploded in perfectly straight lines radiating out from its origin without any deviation, we might just about expect it to collapse back linearly into a blob at some point. The material of the universe is flying past all the other material at very high speeds in countless directions. The effect of gravity is rather like the effect a magnet would have on a ball bearing shooting very rapidly past it. The path of the ball bearing will bend towards the magnet. So it is with gravitational attraction. One great lump of matter will fly past another and each will bend the path of the other towards itself. Mostly one lump will fly past another, their courses deviating slightly towards one another when in close proximity. Only if the speed is slow enough will the larger lump be able to prevent the smaller one flying off by pulling it into a stable orbit. Thereafter the two will settle into a steady rotation at a distance from one another at which the force of attraction is just right to stabilise the orbit. Rather like a person hurling a stone around and around with a sling in a steady rotation, the pull on the sling is just right to keep the stone circling round.

It's this balancing act that drives the orbital motion of the universe: the gravitational force that attracts one object to another also maintains the two in orbit around their mutual centre of mass. Moons circulate around planets, planets circulate around stars and stars swirl around in gigantic galaxies – clusters of thousands or millions of stars. On a smaller scale, satellites orbit around the Earth on just the same principle: their gravitational attraction to the Earth is just enough to keep them turning round in orbit.

Another way of looking at this is to see that satellites (or the Moon) are, in effect, falling towards Earth all the time, without ever getting closer. As mentioned earlier, if they weren't they would just fly off in a straight line at a tangent away from the Earth. To keep turning in an endless gyre, they need to keep falling inwards towards Earth at just the right amount to maintain their orbit. The diagram below shows how a moon (or any orbiting thing) would fly off in the 'v' direction were it not for gravity pulling it inwards along the 'a' direction – and this keeps on happening continuously, forever, all round the orbit (Fig. 12.3).

With this grasp of the basic idea of gravity as a force between any object and any other one, a slew of further questions flowed from the discussion group. 'OK, so gravity gets weaker as you get further away from a planet or a star or whatever,' summarised Julie, 'but does it depend on the shape or the density of a thing?' 'Come to think of it,' interjected Sarah, trying to land a thought she was finding difficult to express, 'where exactly does gravity act? After all, a moon or a planet is

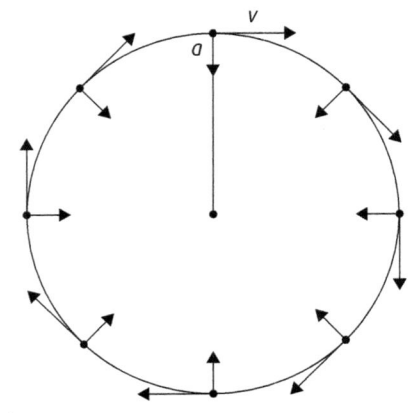

Fig. 12.3 Circular motion

a huge thing – is there some point where the gravity acts?' 'Yes, and what if it's a gas?' added Malcolm, extending the complications even further. 'Do gases have any mass – does gravity affect them too?'

Julie's summary was accurate: gravitational force does get weaker with distance. To be precise, it goes as the 'inverse square', which means that, as we move twice as far away, it gets four times weaker. For a large object, the gravity from every part acts on every other part. I've heard that a mountain can exert a slight pull just because of its physical mass; apparently this can even upset a sensitive pendulum. Local variation in the strength of gravity, though very slight, can be detected with highly sophisticated instruments, and this effect plays a role nowadays in prospecting for metals. On the larger scale, however, it is the average gravitational effect that matters across the whole of a body. When working out the strength of the gravitational attraction between the Moon and the Earth, for example, the force is said to act at the 'centre of gravity' of each object. The centres of gravity of the Earth and the Moon are very close to their geometric centres because they are nearly (but not quite) perfect spheres.

Turning to Malcolm's point about gases, the strength of the force of attraction depends only on the mass, not the density. So a big fluffy foam ball weighing 1 kg would exert the same gravitational pull as a small dense one of the same mass, but made of steel. If this seems to be counter-intuitive, an even more challenging thought is to imagine the gravity associated with something extremely fluffy, such as a nebulous gas. The air, for example, is a mixture of gases comprising zillions of molecules buzzing around all the time. It's not empty space after all. 'Does air have mass?'

asked Mary, suddenly feeling uncertain about something apparently so basic. 'Of course it must do,' interjected Sarah, grasping the point. 'It's made of molecules, each of which has mass.' So gases must indeed experience gravitational forces. It's these very gravitational forces that hold the Earth's atmosphere in place, like a thin blanket around its surface. On the Moon, the gravitational force is too weak to hold on to the molecules of a gas – hence the absence of an atmosphere.

On the stellar and galactic scales the gravitational attraction between gas molecules, weak though it may be, is crucially important. The Sun, like most stars, is just a mass of very hot gas, mainly hydrogen with some helium. The very reason the Sun holds itself together as an orb is because the molecules of gas of which it is composed draw themselves to one another as a sphere. Gases do indeed experience the force of gravity. It not only gives rise to stars such as our Sun, but also holds together all the galaxies and clusters of galaxies that make up the universe as a whole.

Einstein's theory

Persistence in questioning is a useful habit if you are curious about science. Julie and Sonya excel in this respect: rarely do they allow a question to fall by the wayside unanswered. Maybe they harbour a suspicion that science teachers tend to drop questions quietly when they are stumped or, if that fails, subtly divert attention elsewhere. The discussion on gravity was no exception. 'What about the point I raised before, about the curvature of space – I'd heard gravity had something to do with this?' Sonya reiterated.

In the days when I was brought up, long before the explosion in popular science books and television documentaries, the curvature of space was considered a highly esoteric topic, reserved for the closed (and tiny) circle of graduates in physics. So Sonya's line of questioning came as something of a surprise, but nevertheless a welcome one. Perhaps Einstein's 1915 theory of relativity is becoming more widely acknowledged a century on.

It's true that Einstein's new theory of gravitation threw out some of the key concepts that Newton had developed. Revolutionary though this was in an intellectual sense, it's important to add that a revolution such as this in science doesn't necessarily replace the previous regime altogether. It may recast it or revise it or extend it, for example. In the case of gravity, Einstein's new theory reduces beautifully to Newton's old one in the special case of our relatively small, Earth-sized distances and speeds – the scale on which we ordinarily experience the world, and the only ones available to Newton to work with. Einstein's revisions to earlier concepts only make a noticeable difference at the enormous distances of stars and galaxies. Nevertheless these 'small' differences have enabled hugely important discoveries to be made about the very nature and origin of our universe. It was the dramatic measurement of a very slight shift predicted in the apparent position of a distant star that enabled the Cambridge physicist Sir Arthur Eddington to validate Einstein's theory during a solar eclipse in 1919.

In one discussion this historic reference reminded Malcolm, with his wonderful stock of science jokes, of a famous quip: 'Sir Arthur Eddington was once told by an expert on relativity, with whom he was discussing Einstein's general theory of relativity, that only three people in the world really understood it. "Oh, really?" he is said to have responded. "Who's the third?"'

Einstein's revision of the theory doesn't affect how we plan our normal movements here on Earth. But it does recast our concept of the heavens. The very notion of 'force' is abandoned in the new theory. The fundamental idea that the Earth orbits around the Sun, *because* of a gravitational force between the two, is replaced. In its stead is a deeper, more abstract concept that the Earth is simply following its natural course, just as a stone does when it is thrown. The Earth is tracking round in a curved pocket of 'spacetime', a concept introduced by Einstein to reflect his discovery of the interrelatedness of the dimensions of space and time. In what follows the more familiar word 'space' is used to help with visualising the idea of curvature. This pocket is not some kind of a dimple superimposed on the space through which the Earth is travelling.

The profound new concept is that it is space itself – the set of gridlines on the map – that is curved, no longer a straight-line mesh of longitude and latitude. Euclid was wrong; the mesh itself is curved. If we kept on going 'in a straight line', our path would be curved. Ultimately we would arrive back at our starting point.

So, if the very space in which we strive is not linear but curved, what kind of landscape is it? Einstein's answer was that it is shaped by the presence of matter itself. In other words, the Sun and the stars and even the Moon and Earth are creating dimples in the fabric of space. This is very difficult to conceive without imagining these dimples existing somehow in our normal 3D world – like a dip when we sit on a bouncy castle. They don't: they *are* space, not *in* space. Even with its shortcomings this kind of visual metaphor does help us imagine the way heavenly

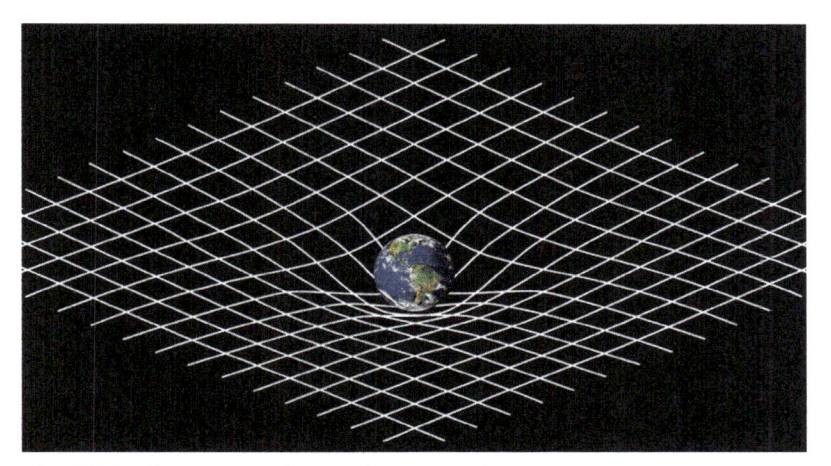

Fig. 12.4 Curvature of spacetime around a mass

bodies move in curved space – the modern explanation of gravity. The diagram offers a representation of what a 2D grid would look like if it were curved in three dimensions by a massive object (Fig. 12.4). If the massive object were the Earth, the Moon or a satellite would orbit round it in the dimple.

The important concept that is impossible actually to visualise is of *three* dimensional space being curved in this way. An interesting implication of this model is about what happens to the path of a beam of light in a curved part of space. Light beams follow the gridlines, so the path will bend slightly near a very massive object, where the gridlines swerve. It was this prediction that Eddington tested when he recorded light from a distant star that just shaved the edge of the Sun. During an eclipse this star could actually be seen in daylight, and it was found to be displaced, as predicted, from where it was expected to be.

The implications of either of the two gravitational theories – Newton's or Einstein's – are profound for the universe as a whole. On the largest scale, if all the matter in the universe is attracted to all the other matter, wouldn't the universe eventually just collapse inwards? Well, as discussed earlier, it just isn't doing this. In fact, as Edwin Hubble discovered in the last century, it is actually *expanding*. In even more recent times it has been found to be not only expanding, but also to be getting faster and faster as it does so. This has become an active area for research today, as cosmologists try to establish just how far the explosive energy of the original Big Bang will continue to push an expanding universe against the gravitational tendency for matter to clump together.

Conclusion

Terry's trip to Blackpool had inspired a simple and unassuming question about its beach. Watching the ebb and flow of ocean tides is an experience open to any of us, not in itself an exotic scientific phenomenon. Yet the questions it provoked have led into some of the most profound thoughts of mankind. By reflecting on the monthly rhythm of the tides, we have turned to the influence of the Moon and thence to the nature of gravity itself. Explanations of this have, in turn, thrown up profound ideas about how distant bodies are able to interact with one another. A simple question from everyday life has opened up thinking about fundamental scientific concepts. But perhaps this discussion has left Terry, with his original inquiry about the beach, somewhat short-changed. 'I still don't see why the sea is so far away at Blackpool' he might well have retorted. 'Is it to do with the flatness of the beach? Maybe we need to talk about geology and sedimentation?' Perhaps we should, but let's leave that for another day.

Questions about the nature of the universe are just as likely to inspire discussion as ones about electricity in the kitchen. Fascination with scientific ideas seems to derive from both the exotic and the everyday aspects of our lives. What is interesting, when discussion is allowed to run freely, is the way in which the two are so often intertwined. On one occasion, a discussion triggered off by thoughts about the origin of our universe led into an exploration of one of the most pervasive of all scientific concepts: energy. Varied as its everyday connotations are, the concept of energy has been refined in science over the centuries to become one of the key ideas linking the many branches of science and technology. Its importance is reflected in its relevance to so many areas of public concern today: from health and fitness to the environment and even international relations. The concept of energy is explored over the next three chapters.

13
Energy

'OK, but what was there *before* all the atoms and molecules?' We were talking about the origins of the universe in a wine bar, as you do. We had been working our way backwards from the universe we know about today, with its stars and planets, meteorites and asteroids, towards its earliest moments, just after the Big Bang. You start thinking about how atoms came into existence in the first place – have they always been there? Then you go back even further. What about the particles that make up atoms – the protons, neutrons and electrons – how did they arise? Continuing on, you get back to what we now see as the fundamental components of matter – the quarks that go to make up protons and neutrons. In each of these stages one form of matter aggregates together to form the next, from quark to proton and neutron to atom. Yet this wasn't really the issue. Our questioner had accepted this account of the way the material universe had evolved and was now addressing an even more challenging concept. There must have been an even earlier stage before stuff of any kind existed – atom, proton or quark. What was there before matter?

This is a profound question, demanding imagery that is more or less impossible to summon up, but in scientific terms the answer is really quite simple to state – there was just pure energy in the time before the matter that makes up today's universe came into existence. Fair enough, we might say: energy, we've heard of that – at least the universe was kicked off by something familiar, not some strange, incomprehensible thing. Energy is something we know about; we use it every day, our lives depend on it, we even receive a bill for it every quarter! But is energy really such a familiar concept? Do we understand its meaning in any depth? To quote from the inevitable question raised in the wine bar discussion: 'What actually *is* energy?'

Connotations

This simple question, thrown up by thoughts about the universe, led into deep and unpredictable discussion about some very down-to-earth aspects of our daily lives, from boiling milk to keeping fit. Simple though the question is to pose, working out what it means is not so easy – which is strange for a word that features so often in everyday conversation.

'It's something that can be felt but is intangible – like the heat of the Sun,' suggested Celia, always good at finding a link to everyday experience. 'Isn't it is a force of some kind?' ventured Rosie, trying another tack. 'It's to do with physical fitness and wellbeing, surely – whatever it is you never seem to have enough of it,' Jean claimed wryly. This observation led on to a further everyday association: the energy in food. Calories are a big talking point today as many people struggle to control their diet and regulate their weight. If you have the patience (and eyesight) to read the small print on many food labels you'll see the word 'energy', setting out how many calories are to be found in every 100 grams of the food.

Of course, it's not only physical things, such as sunshine and food, that the word brings to mind. Another area of meaning is summed up by a comment from Michelle: 'It's when you feel something about somebody – an intuition. People give out positive or negative energy vibes.' This more spiritual association leads on to the cultural context in which the word is used: as Sonya put it, 'Do we mean a Western definition? In my country of origin it's common to talk of energy in this way.' Indeed in traditional Chinese medicine the concept of 'chi', which is considered to flow through the body along channels known as meridians, is routinely translated as 'energy'. The word energy is understood in many different ways – it all depends on the context in which it is used. It was not until the mid-nineteenth century that the word was commandeered for its specific role in science. Originally a Greek word associated with activity, 'energy' came to be used to denote power or vigour in, for example, action and speech. It is not unusual for everyday words with a range of meanings to be adopted in science to denote something specific. This plundering can lead to confusion, if scientists take for granted the meanings they have been trained to adopt, while everyone else understands the terms in their vernacular sense.

Scientific concept

Energy is commonly defined in science as 'the capacity to do work'. This is routinely taught in schools, but it is not obvious what it truly implies. Let's take the opportunity to look more closely at the phrase. The word 'capacity' itself has many connotations: the maximum something can contain or the ability to do something, for example. It's the latter meaning that is behind the concept of energy. It suggests that energy is not a substance, nor any kind of tangible or visible thing. Instead the word describes something abstract: the *potential* to do something. Thus a closed jack-in-the-box has the potential to spring open, a sugary drink the potential to fire up your muscles, a poised hammer the potential to drive in a nail. In each case, this abstraction – energy – seems to get stored up in some kind of physical system. A useful analogy is with money, also an important but abstract entity. Normally nowadays money has no substantive physical presence, but is merely represented by figures on a statement or words on a bank note – yet it has the capacity to do an awful lot for us when released!

The other part of the definition is about 'doing work'. This tells us about what happens when the 'capacity to do something' is actually realised. When energy is released, the textbooks tell us it is 'able to do work'. But is this strictly correct explanation really helpful? As so often happens in learning science, one definition simply pitchforks you into a further one; self-referring cycles of definition can seem as maddening as a drawing by Max Escher.

'Work' can be explained fairly simply because the scientific meaning is not too far removed from the everyday one. Work is done when a force is applied to something as it moves. So, for example, when your legs push bicycle pedals against the friction of the tyres on the road, they are doing work. The energy you have in your legs has the capacity to do work against friction. Examples abound, in trains and planes, machines and muscles – forces are moving things everywhere.

The word energy was first used in a scientific context in 1807 by the English investigator Thomas Young, and was given its modern scientific meaning by William Thomson (later known as Lord Kelvin) in 1852. Yet in an 1847 lecture, even the leading physicist (and brewer) James Joule still employed the more archaic phrase 'living force' (*vis viva*) to describe what we now call energy. These men were living in an age of unprecedented growth in manufacturing and industrial processes generally. New ideas and inventions abounded: science and engineering were proving not just interesting concepts, but vital elements of industrial growth. In fact scientific understanding of what we now call energy developed hand in hand with economic development. It was not so much disinterested scientists, dwelling philosophically on the concept of energy and its various transformations, who stimulated experiment and theory; progress was rather driven by the owners of mines and railways, anxious to increase the efficiency of their steam engines. The realisation that heat generated by a coal fire could result in movement of a heavy locomotive or pump led to investigations of the link between heat and mechanical movement. The older concept of heat as a kind of fluid inside materials (dubbed 'caloric') was abandoned in favour of the concept of it as energy – something that could change its form, as heat was converted into mechanical movement, but whose total amount was always conserved.

'Ah', interrupted Mary at this point. '"Energy is neither created nor destroyed" – I remember learning that at school. So that's what it meant!' She was right: the chemical energy released when coal combines with oxygen in the burning process reappears as energy in the form of heat. This in turn creates the steam that drives a piston, whose motion converts the energy into a mechanical form. When all the energy is accounted for, whether it resides in the motion of the locomotive or is lost as heat to the surrounding air or in the frictional rubbing of machine parts, the total sum remains the same as the amount that was released in the original combustion of coal with oxygen. Although today this concept of the 'conservation of energy' seems straightforward and almost self-evident, it was not arrived at through abstract reasoning but through meticulous measurements conducted in carefully controlled laboratory experiments in the mid-nineteenth century.

The concept of energy as a kind of interchangeable currency rather than a physical substance was a major step forward, and one that seemed counterintuitive at the time. We might well wonder what happens if we get up from a chair, eat some food, go for a run and end up back in the chair we started from – was energy really conserved? It seems to have just gone, dissipated. It is only when all the heat energy that we have given up to the surrounding air, as we run and digest, is taken into account that the total sum expended is seen to balance the energy consumed.

Talking of food reminds us of one of the commonest references to energy in our daily lives: the way it appears on food labels. What does it mean, what is a kcal? We are well aware that food contains energy – after all, we talk of needing an energy burst from glucose when we are exhausted. Yet where is the energy in a chocolate bar? How is it stored and how does it get released into our bodies? Even more perplexing, how on earth is this linked to the energy in steam locomotives and electricity supplies?

Unrelated though these questions may sound, the answers reveal a surprising unity. Ultimately the energy expressed in each of these contexts has its origin in a similar place: the arrangement of atoms within molecules. Foods, plants, coal, oil – all these substances are composed of molecules made up of atoms. These molecules have been forged in chemical reactions during which the atoms of the combining molecules were re-arranged and locked together – rather as the character in a jack-in-the-box is pushed into the box and the lid latched. In other words energy gets stored up when things are pushed together against a resistance and then firmly locked together. The energy in food and fuels is associated with the way in which atoms hold together within molecules, much as the energy of a jack-in-the-box is associated with the compressed spring kept in place by the latched lid.

If we want to consider an even deeper level of explanation, we might ask where the energy came from to put the atoms together into a molecule in the first place. For our jack-in-the-box analogy, this would be the equivalent of the human energy that compressed the spring as the lid was closed in the first place. In the case of food, coal or oil the ultimate source of this energy isn't to be found anywhere at all on Earth. As you might have guessed, it all derives from the Sun, the source that drives the chemical reactions in plants as they grow. Plants are harvested for food, grazed for meat and, millennia ago, became the raw material for the coal beds and oil fields that power our locomotives and generators today.

'All very interesting' was the general reaction as this explanation was unfolding in one discussion group, 'but how do we get from the idea of the arrangement of atoms in molecules to the sensation of energy we experience in our bodies – or lack of it when we feel exhausted? How does the energy in a cheese sandwich or glass of beer work its way into our system and get us going?' As we eat and drink, wonderful as the experience can be, all we are doing from a biological point of view is getting plenty of the right kind of molecules down into our digestive apparatus. There, assailed by enzymes in a bath of highly concentrated hydrochloric acid, these energy-laden molecules are gradually broken down into smaller ones capable of passing through the membranes that line our gut. Being made of large molecules themselves, these gut membranes are quite capable of permitting the smaller molecules from our food to pass through and enter the blood vessels close by. Thus high energy molecules enter our bloodstream and, by this means, circulate throughout all parts of our bodies. Almost every cell of the body contains places (called mitochondria) where the energy contained in these nutrient molecules is transferred to a standard energy carrier that is then used throughout the body. The way in which this energy carrier, known as ATP, goes on to get you moving and thinking is discussed in chapter 15, Energy for Life. If you exert yourself, physically or mentally, without adequate nourishment, it's no wonder that you feel exhausted. In a colloquial way, we complain of having no energy and, by and large, that's just right: energy is indeed what we are short of.

Conclusion

Today the word energy has so many connotations in everyday life that our thoughts about it have ranged widely – over physical things such as forces and movement, biological aspects, for instance the food we eat, and also more personal things to do with our culture and beliefs about people and our bodies. Not surprisingly the scientific concept of energy appears rather narrowly defined in relation to these broader notions. Indeed it is; this is how science proceeds, by establishing a universal definition that enables measurements to be made and relationships to be quantified.

This more precise use of the word 'energy' has given rise to countless experiments and insights that have gradually extended use of the concept to previously disparate areas of knowledge: heat, mechanics, light, electricity and, more recently, nutrition and physiology. It is now central to our understanding of a host of contemporary global issues, from climate change and nuclear energy to feeding the hungry and combating disease.

In the next chapter we build on the concept of energy outlined here by considering what happens as energy moves around. It's particularly relevant to our everyday lives when we consider the flow of heat energy through the walls and windows of our homes and the vital energy transformations that drive our ovens and refrigerators.

14
Energy on the Move

'Where does all the energy go when a cup of tea cools down?' asked Sonya, inspired by the energy theory outlined in the previous discussion. Her question picked up on the idea of conservation: the idea that energy never actually disappears, but simply shifts from one apparent form to another through some physical process. In the case of a cup of tea, it's easy to see where the energy comes *from* when it's made. Passing electricity through the coil of a kettle or burning gas under a pan on a hob are both ways of releasing energy, which in turn raises the temperature of the tea. This energy ends up in the form of internal energy or heat in the hot tea. Fair enough; but where does it go next?

Clearly the tea will soon cool down. It's an obvious fact that its temperature will gradually drop until it eventually reaches the temperature of the surroundings. If we were to put our faces close to the outer surface of a cup of tea while it's cooling, we'd notice the air around it gets warm. This gives us a useful clue about our question: heat energy is clearly transferring away from the hot cup into its surroundings. If we were very observant, we might even notice that while the tea is still hot it transfers its heat quickly, but as it reaches lower temperatures it seems to cool more slowly, ultimately keeping slightly warm for quite a long time. I remember my school physics teacher starting a lesson once by quizzing us on whether it's best to add the milk first or at the end if we're in a hurry to drink a hot cup of tea. It's useful to think this one through if you're ever caught in a station cafe with only a minute to go before your train arrives! This question turned out to be just a prelude to learning about Newton's law of cooling.

The way in which energy dissipates may even bring back memories of another mantra from school science lessons: the trinity of 'conduction, convection and radiation'. These three terms succinctly describe the various ways in which heat energy can move about. The first is the way it simply transfers from one molecule of a substance to the next. In the case of a hot cup of tea, energy is shifting sideways from the molecules of the cup to the adjacent molecules of the air. The second, with its hint of the Latin word *vector* ('one who carries or conveys'), describes how it is physically carried away in the current of warmer air as the latter becomes less dense around the cup and rises. The third, as its name implies, describes energy being radiated away in all directions, carried by invisible waves that simply pass through the surrounding air. An interesting challenge is to think what would happen were you ever to be caught with an overly hot cup of tea on the Moon. There's no atmosphere at all up there, so it would still cool down as the heat energy gradually radiates away. Yet with no air around to enable conduction or convection, the process would surely take a lot longer than on Earth.

These ideas describe the way heat energy travels and give some sense of the speed with which it moves. 'That's a good point,' acknowledged Wai Ling, an occasional member of the group who was a potter. 'You can heat things up at different rates. Raku firing, for example, is slower than a normal kiln and produces different effects on the glaze.' Clearly the pace at which heat transfers can affect the way it alters substances. Talk of the speed of heating inspired Lucy to recount 'a worrying experience in my bathroom'. She recalled a day when an unexpected burning smell had wafted down her stairs. A towel on a wooden rack was smouldering away mysteriously, with no apparent cause. 'Then I realised the Sun happened to be shining directly on to a curved shaving mirror on a shelf, in such a way that it directed the rays on to the towel and focused them on to a small spot.' Heat radiation from the Sun must have been intense enough to raise the temperature of the towel to burning point – a frightening experience, and dangerous had the house been unoccupied, but nevertheless a fine introduction to energy transfer.

Heat flow

The common idea behind these varied experiences is that heat energy moves. In most processes we experience it is continually moving: indeed it simply *has* to move. Even when we speak about storing heat, for example in a hot water tank, some heat energy is always leaking away to the surroundings. An obvious point that we learn intuitively is that heat energy flows when there is a difference of temperature, for instance from a cup of hot tea to the surrounding air, or from a gas burner on a hob to the base of saucepan. Equally obvious, but of profound importance theoretically, is the observation that heat energy will only flow from places of higher temperature to ones at a lower temperature. Thus if a cold spoon is placed in hot water heat will not flow from the spoon to the water, only the other way round.

Common experience also tells us something more about the flow of heat: not only does it flow from hotter to cooler places, but it also moves faster when the difference of temperature is greater. I notice this particularly on cold but sunny days in winter. During the day, while the Sun is heating the air outside, the central heating seems to cope perfectly well, but as evening comes on and the outside air drops rapidly to a much lower temperature than the room, the outward flow of heat through the windows, walls and door gaps speeds up dramatically, making the room suddenly cooler. All the more reason to bring in the insulation specialist to reduce the conduction and convection of heat from your house by packing the walls with foam and doubling up the glazing.

So it's a difference of temperature that causes heat energy to flow. It's useful to draw an analogy with the journey of a river. Water always flows downhill, never up, and it's the difference in height that drives its progress – gushing down a steep mountain gulley, for example, but only meandering lazily through the flat plains of an estuary. Another comparison is with the circulation of blood in the body; in this case it's the difference in pressure, created by the pumping action of the heart muscles, that drives the flow.

The nature of heat energy

'Does cold flow?' was an interesting query from Sarah as this story of heat flow began to unfold. It's a good question, highly relevant as I write on a typical winter's day in Britain. Cold seems to be rushing in through the tiny gaps around the door frames and stealing in from the exposed walls of the bathroom extension. It certainly feels as though cold flows from a colder to a hotter body as surely as heat flows the other way round. This imagery provokes a deeper thought: if heat is a form of energy, what exactly is cold? Is it a kind of negative energy? Is it some ethereal substance?

These ideas about the nature of heat (and cold) were studied thoroughly in the nineteenth century. Up until the middle of the century the prevailing theory held that heat was a kind of invisible substance that flowed through things, from hotter to cooler regions. Many experiments were undertaken to pin down the nature of this supposed substance, which was even given a name: 'caloric'. Eventually the whole idea was abandoned as it began to be understood that heat was not a substance at all, but simply a manifestation of the agitated motion of the things that make up all matter: the molecules. So when Mary once asked in a discussion 'Is cold a thing – or is it simply a lack of heat?', she was definitely on to something. It turns out that heat is not really a thing at all. Not surprisingly 'cold' isn't really a thing either. Both sensations are caused by molecules jiggling about – the more they jiggle, the higher the temperature and the greater their energy.

With this image in mind, we can now understand the logic of heat flow. The agitated motion of molecules in a high temperature region spreads over time to the less agitated molecules in surrounding regions that are cooler. It's this gradual transfer of the energy of agitation of molecules that we sense as heat flowing. Thus when the cold of winter seems to be creeping in through our window panes, it's really the other way round: the agitated motion of the molecules in our warm rooms is being passed on continuously to the neighbouring molecules of the cooler glass and then on to the even cooler air outside. No actual substance moves, but the energy of the moving molecules passes along the chain of molecules to the less energetic ones outside. As a result our rooms feel cooler near a window as the molecules inside become less agitated, giving up some of their energy to their neighbours. With this image in mind, the less energetic molecules outside aren't really passing on their 'coldness' into the warm room; instead they are drawing out the heat energy from the room. So no, cold doesn't flow.

To some in the group this revelation about heat flow raised new uncertainties. 'Are you really saying that heat flows from any temperature to any lower one?' asked Sarah, thinking imaginatively. 'It seems strange to say something at a negative temperature, say $-10°$, can pass on heat to an even colder thing at say $-20°$?' 'Yes' added Michelle, 'What happens if you are in the Arctic and you expose an ice cube at $-10°$ to the surrounding air at $-20°$? Does it heat the air up?' The answer is yes, it does: an ice cube does contain a certain amount of heat energy and it can indeed pass this on to an even cooler thing. In a similar way a cool bath still contains heat energy, even though there's less of it than in the same bath when hot. In fact it is even possible to extract the heat energy from things that are merely warm, such as used bath water or simply the air that surrounds us. This is in effect what a refrigerator does – it takes heat energy out of the food compartment and pumps it out into the room through the fins at the back. One way of heating a home is to use a device known as a heat pump, which acts like a refrigerator working back to front. This reversed refrigerator is placed in the external wall of a house in such a way that the hot tubes we'd normally find on the back face instead into our homes, while the cold chamber is outside in the surrounding air. The bountiful air outside is cooled a little and the room inside is heated – a fridge in reverse to heat your home.

'OK, so there's a certain amount of heat energy even in cold things like ice cubes. It seems very strange, but I accept that,' Julie declared graciously. Sarah joined in at this point, checking her understanding with a probing query: 'It's just that there is less heat energy in a cool bath than in a hot one. So are we saying that if I lit a match and held it against a glacier long enough it would melt it?' Another interesting question arising from the group – and one that gets to the heart of a big confusion surrounding our everyday understanding of heat. When are we talking about *heat* and when are we talking about *temperature*? The English language is absolutely no help on this – it only reinforces the confusion. After all, one minute we say, 'brrr, it's cold out there' (using 'cold' as an adjective), and the next 'quickly, shut that door, you're letting in the cold' (using 'cold' as a noun). Talking of language, scientists themselves now refer to 'internal energy' rather than heat energy partly because of the confusion we've just discussed. Linguistically, it seems odd to speak of the 'heat' in an ice cube, so scientists often talk about greater or lesser amounts of internal energy instead. For simplicity, however, we will keep the more familiar phrase 'heat energy' here.

So let's consider whether a match can melt a glacier. We know that heat energy flows away from an object at high temperature, such as a match flame, and towards an object at lower temperature, for example a glacier. So yes is the answer: energy will flow into the glacier and, when it gets there, it will be absorbed. As the energy flows into the glacier, the temperature in the neighbourhood will rise and a little bit of ice will be melted. Unfortunately the match will soon burn out, so it will not have much impact on the glacier as a whole. The principle stands, however: if you go on transferring energy into the ice and making sure it gets to all parts of the glacier, it will eventually melt – though it will take some time, and use up quite a lot of matches too! Heat energy from the increasingly warm seas and surrounding air is doing this right now, in fact, as one aspect of the climate change we are experiencing.

Heat, temperature and thermal conductivity

Putting two of our images together, we can now see that a warm bath may contain more heat *energy* than a white hot match flame, but the *temperature* of the bath is very much lower than the flame. If we applied the white hot match to the side of the bath to heat it up, we'd barely notice the rise in temperature. This image of a match flame and a warm bath helps us to think more clearly about the different meanings of the words 'heat' and 'temperature'. The word 'heat' is associated with the *quantity* of energy involved and the word 'temperature' with the *quality* of it. A thousand units of low quality heat energy in a bath may warm you up, but the same amount of high quality energy in a flame can ignite a fire.

Talking of temperature, Patrick remembered clearly from his first visit to Italy something about floors found in houses, churches and public buildings there. They are often made of a hard stone such as marble or large ceramic tiles – and are not often carpeted. Patrick was struck by the cold sensation in the soles of his feet as he got out of bed in his *pensione*. Accustomed as he was to carpets and wooden floors, he soon realised that a cool floor was highly desirable in a country where summer's baking heat can be uncomfortable. Yet no sooner had this idea struck than a rather confusing thought came into his head. Surely, under normal conditions, everything in the room should have settled to roughly the same temperature? If we had a cold floor and warm air, the air would simply warm up the floor till they were at the same temperature. 'Is a tiled floor really colder than a carpeted one?' he queried.

As soon as we think about it, we see that this cannot be. All the objects in a room must be at the same temperature, provided things are settled. Of course if the Sun is beating in through a window, that area might be at a slightly higher temperature, as may the air around a hot radiator. In the absence of a source of heat like this, however, everything should settle to the same 'room temperature'. But Patrick's bare feet were undoubtedly feeling colder on the tiles than they did on his carpets at home. What was happening? Strange as it may seem, it was not actually a cooler temperature that his feet were sensing at all – it was the flow of heat energy from his warm-blooded feet draining away rapidly through the tiles. Had he stepped out on to a carpet instead, the heat wouldn't have drained away so fast. It's not really so mysterious– our feet are just sensing the loss of heat energy rather than the actual temperature. Feet aren't thermometers, after all. Quite sensibly our bodies are detecting how rapidly valuable heat energy is leaking away – that's the real threat to our wellbeing. This clears up another area of potential confusion.

In summary: heat is a form of energy that flows when there is a difference of temperature. Our feet are at a higher temperature than the surroundings, so heat tends to flow from our feet into the floor. It's not advisable for our bodies to lose energy too rapidly, so sensors in our bodies alert us to how quickly heat is draining away. The key is the speed at which energy is flowing away. Ceramic tiles conduct heat well, so contact with a tiled floor means that our body heat drains away more rapidly than with a wooden or carpeted floor. Both wood and carpet are much poorer conductors of heat.

Conclusion

With this insight into Italian floor materials, we now are able to separate out some of the key concepts in the area of heat that are so easily confused. The word 'heat' refers to a form of energy that flows; there can be a lot of it, as in a warm bath, or only a little, as in a cup of hot water. Temperature, on the other hand, tells us which way heat energy will flow. If we pour a cup of hot water into a warm bath, the heat will flow from the cup into the bath. Conductivity tells us how quickly heat energy will flow through any given material. Touch a hot oven dish with bare hands, for example, and you'll soon feel the flow of heat; touch it with insulating oven gloves on and you won't feel the heat for much longer.

As a coda to the interesting story of heat flow, we can see just how pertinent the topic is in everyday life. If we try to reduce our heating bills, we know we are encouraged to install better insulation. What does this do but slow down the flow of heat energy? We use double glazing to introduce a thin layer of air between two panes of glass, as air is a poor conductor of heat. Insulating wool laid in a loft traps air in its spaces, with very similar effect. Preventing heat loss is one of the key themes in how humankind might reduce its demand for energy, a central issue as we strive to reduce carbon emissions to minimise the risks of climate change. Our future depends on capturing more heat energy directly from the Sun and the winds it creates; survival depends on maintaining temperatures within the band that fosters temperate weather and the flourishing of crops, animals and ourselves.

Discussions often come round to the unifying concept of energy even when they have utterly different starting points. The preceding two chapters began with the origin of the universe, with its unimaginably intense concentration of primordial energy, then moved on to its mechanical form in hammers and bicycles, its chemical form in coal and food and, finally, to its manifestation as heat, in warm baths and white hot flames. Understanding the way in which energy moves from one form to another, whether in the cascading water of a mountain stream or burning oil of a power station, is central to almost all branches of science.

The life sciences are no exception. The basic processes of digestion and respiration underpinning our very existence have evolved to ensure we have the energy needed to move around, keep warm and, above all, to think. The discussion retold in the next chapter focuses on the body's energy system. It didn't begin that way, however – far from it. At the time of the discussion in 2015 newspapers were filled with headlines about so-called 'three-parent babies' – and the science behind this advance in fertility treatment was the starting point.

15
Energy for Life

The mitochondrion had hit the news. What on earth was this unfamiliar part of the body? Was it one of those bits whose name we struggled to memorise for biology exams long ago? Who would have thought that such a remote word would become a public sensation overnight? The cause of this rapid rise to fame was the discovery in 2014 that a healthy embryo could be created for women with a terrible genetic condition if the cells of three people were used rather than the usual two. The lurid headlines played on the emotional consequences of 'three-parent children' and the potential legal ramifications. The story beneath the headlines provided the trigger for a fascinating exploration of how the body uses energy – a story crossing the boundaries of chemistry, biology and physics (not to mention nutrition, genetics, embryology and more).

Helen asked the first and most fundamental question: 'What exactly do the mitochondria do?' It is indeed one of those little bits you probably had to learn about in biology lessons. It is also possible that you have forgotten what you learned at the time. Mitochondria are one of the very small pieces of apparatus that float around inside a cell. Cells are small (ranging in size from one-tenth to one-hundredth of a millimetre), but like most biological structures they are themselves made up of parts. The most amazing thing about the cells in our bodies is that they all have the same basic structure, whether they form muscles, nerves, skin or bones. They have flexible walls (known as membranes) that separate their watery internal contents from the watery external world. The membrane consists of two parallel layers of long, thin molecules called phospholipids. Each molecule has one end that mixes easily with water and another end which is oily and does not. The diagram shows a section of membrane (Fig. 15.1). The long lipid molecules lie next to each other in two layers. The molecules are orientated so that the ends which mix with water (shown as red in the diagram) are on both

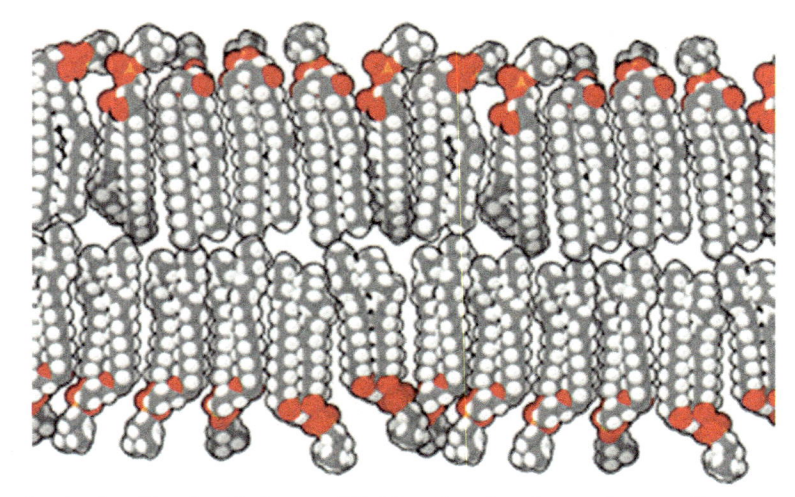

Fig. 15.1 The double layer of lipid molecules that makes a cell membrane

the outer and the inner surfaces of the cell (the upper and lower parts of the diagram), while the middle part of the membrane is essentially oil-loving. This enables the membrane to tolerate the watery environments both inside and outside the cell, while repelling water within itself. It makes the cell watertight.

This compartmentalisation of cells – separating the inside from the outside – is essential to enable cells to carry out their roles. Cells need to do various kinds of things, such as storing insulin or carrying around oxygen, and they need to be self-contained. The cellular structure also reflects the way in which the organisms grow and develop over time, by multiplication of this basic repeatable unit. In contrast to manmade compartments such as prison cells or gym lockers, designed to isolate their contents completely, biological cells are able to admit substances in and out through their membranes. This is how they survive over the long term, by allowing the materials they need to pass through and by expelling those they don't require as waste. In this way small molecules vital to life, such as hormones and energy-rich compounds, are able to get inside cells and produce their beneficial effects.

Inside the cell membrane are several pieces of apparatus – known collectively as organelles. One of these, the nucleus, stores genetic information in the giant DNA molecule. Another, called the ribosome, enables information from the DNA to be transcribed into the various building materials and tools the body needs, in the shape of various kinds of protein. However, the particular organelles involved in the so-called 'three-parent baby' story are less well-known ones: mitochondria. They are rather like miniature cells within a cell, surrounded by little membranes of their own. Their special role is to convert energy that is released when molecules from the food we eat combine with molecules from the oxygen we breathe in. Mitochondria make this energy available for use in the cell. They do this by transferring the energy through a series of chemical reactions to specialised molecules, which then carry the energy to where it's needed. These molecules, known by the simple acronym ATP (an abbreviation for Adenosine Tri – Phosphate), supply the energy needed for activity in the muscles, the brain and other parts of the body.

'Can we pause a moment here, Andrew? I don't think I'm the only one who is unsure about what energy actually is. I know we've talked about it before but can you remind us?' This plaintive cry came from Stephanie, a psychotherapist who had in fact studied science at 'A' level many years ago. It was a good question at the time and gives an excuse to remind ourselves of what was said in chapter 13.

Energy is essentially an abstract concept, not something substantive. In physics it's defined as 'the capacity to do work', which means that when a system changes from a higher to a lower state of energy things happen, work gets done. Energy keeps circulating round, changing its form but remaining unaltered in overall quantity.

The energy stored in chemicals is released when the atoms are rearranged during chemical reactions. The amount of energy released in this way can be measured. In fact you routinely hear about this in connection with the number of calories in foods. A calorie is simply a measure of energy (an old-fashioned one, in fact). The Calorie with a capital C is used for foods to denote 1,000 calories as the latter is an inconveniently small unit. When a pot of yogurt is labelled as having 120 Calories per 100g, it shows the total amount of energy released in this way for every 100 grams of the yogurt (roughly half a typical pot). After consumption the molecules in our food – the various proteins, fats, sugars and so on – are broken down in the digestive system into a limited number of basic molecules, mainly a type of sugar molecule known as glucose. The molecules of glucose are absorbed into the bloodstream from the small intestine. This glucose passes into the various cells of the body and on into the mitochondria inside each cell. Here energy is released when the molecules of glucose combine with oxygen.

Of course the amounts of energy involved in all these activities are absolutely microscopic compared to the energy used in boiling a kettle or driving a motor car. A teaspoon of yogurt may contain roughly 10 Calories or 42,000 joules of energy (the standard unit of energy). The amount of energy released when a single ATP molecule acts on a muscle is miniscule, about 0.00000000000000000000051 joules. The energy economy of a human body consists of zillions of miniscule transactions taking place every time a muscle quivers or your skin cools down a fraction. Each involves just millionths of billionths of billionths of joules of energy. Added together, the total energy used every second by all the chemical reactions in your body approximates to a single 60 watt light bulb. That means it's using up 60 joules of energy every hour. So humans are quite low-energy beings really, but of course highly efficient in the way they use it.

It's at a point such as this that a discussion group begins to realise that the various twists and turns of its conversation, interesting though they may be, have taken it a long way from the original issue. Someone often interrupts to remind the group of its starting point. On this occasion the task fell to Helen. 'Andrew, it's all very fascinating, this story of how energy is used in our bodies, but what's it got to do with the "three-parent babies" we were talking about in the beginning?' she asked.

In fact it has lot to do with it, especially for the children suffering from mitochondrial disorders, and for their parents. Put bluntly, if the mitochondria don't work properly all the energy-consuming systems of your body are at risk: that means loss of muscle coordination, damage to hearing and sight, liver and kidney disease, as well as a host of other symptoms. Until recently there was no cure for this condition. The problem is that mitochondria uniquely depend on their very own supply of mitochondrial DNA to function. The other parts of a cell use the main stock of DNA that is wrapped up and stored in the nucleus, a different organelle within a cell. The amount of DNA in the mitochondria is, however, minute in comparison to that in the nucleus. In rare cases the DNA in the mother's mitochondria are defective; this will be passed on to her children, whose mitochondria will then have the same problem. It is an inherited genetic defect, passed through the mother. In the new procedure, first tested in 2014, the egg to be used to create an embryo is taken not from the biological mother, but from a donor whose mitochondria are healthy. In order that the embryo's DNA comes from the biological mother, however, the nucleus of the donated egg cell – with its third-party DNA – is first removed, and the nucleus of the biological mother's egg is then inserted into the donor cell in its place. It is this that contains the vital DNA that will determine the characteristics of the new baby. The new hybrid cell contains the mother's DNA inside a healthy donated cell. It is this that is then fertilised by the biological father.

Sally was keen to pick up on an earlier point about energy in the body. 'You say that the energy stored in molecules is released when the atoms that make up the molecules get rearranged in a reaction. How does this work?' This intriguing question may have arisen in the context of the body, but it opens up a fundamental idea about energy that applies in all its contexts, whether biological or physical.

Systems in general tend to shift spontaneously from a position of high energy to a lower one, but not the other way round. For example, streams at the top of a mountain will flow spontaneously down to lower ground – from a higher to a lower level of gravitational energy; they won't flow uphill spontaneously. Similarly a battery will drive a mobile phone for a while, as the chemicals in it gradually react, shifting from a higher to a lower state of energy. In the right circumstances, however, energy can be held in a higher state instead of flowing spontaneously; it is then ready to be released when needed. This is what a hydroelectric dam does up a mountain, holding the water at a higher energy level until the energy is required. It's also what a battery does when the device that contains it is switched off, storing its energy until the device is switched on again.

Of course, energy that is stored in these ways has to be put there in the first place by some means. A mountain dam reservoir is fed by streams falling from even higher energy levels. A battery has to be recharged from the mains (or replaced). The same happens for food acting as a store of energy. The molecules that make up our food come from plants that acquired their energy from chemical reactions powered by the Sun while they were growing. In the process high energy molecules were created, mainly carbohydrates, fats and proteins. It's the energy from the Sun that was used in the first place to forge the high energy arrangement of atoms in these food molecules.

The remarkable virtue of the notion of energy is that the self-same concepts about it apply in every conceivable context. From the human body to the interior of stars, from electrical supply to the growth of plants, we can imagine energy being stored, transformed and released in an infinite variety of physical processes. This is indeed an act of imagination, given that energy is not a tangible substance. It is one of the great abstractions, a measurable quantity we can never visualise, but whose endless transformations we can account for in the most minute detail – from the tiny fraction of one joule of energy each time a mitochondrion powers up an ATP molecule to the 10,000,000,000,000,000,000,000,000,000 joules produced each day by the Sun. We owe a lot to the brilliant engineers and scientists from the age of steam who developed the modern concept of energy so fundamental in today's world.

Another form of energy, one that for many people is as mysterious as it is helpful, is electrical energy. Electricity is invisible, silent and essential for almost everything we do. It's so powerful that it can kill in a flash, and so responsive it senses the flick of a finger on a screen. In chapter 16 some of the basic concepts in electricity are explained as they cropped up in one, rather exhausting, discussion.

16
Electricity

'Why do batteries go flat?' asked Sonya. This simple question launched an avalanche of further ones and began a long series of discussions about electricity. What's inside a battery? Why are there three holes in a socket? Why doesn't electricity leak out? What are volts and amps and watts? What is static electricity? How is electricity made? Little wonder there are so many questions, given the major role electricity plays in almost every aspect of modern life. There is the obvious setting of the household with its various devices and circuits. Then there's the various types of communications apparatus – radios, TV, mobile phones – and all the electrical parts in cars, planes and ships too. Many biological systems are also electrical – that's how nerves send their signals and muscles make their movements. The cosmos itself is electrical with its cosmic rays, lightning strikes and the Aurora Borealis (or Northern Lights). All these manifestations to contemplate before we even enter into what electricity actually is and where it comes from!

Amps and volts

Where should we begin? Perhaps the best way would be to establish what we already know from common conversation and everyday life. 'Well, let's see,' said Sarah, rising to the challenge. 'I suppose it must be something that flows. After all, we talk about electric current just like the water in a river.' It does indeed flow; we know electricity has to move to get from the power station where it is produced to our homes, or from the battery to the screen of our mobile phones. Roughly speaking, then, an electric current is a flow of electricity even as a river current is a flow of water (though there are differences). The amount of water that flows through the River Thames at London Bridge is around

70,000 litres every second on average. This is a measure of the rate at which the water flows, and that's basically what amps are about – they tell us how much electrical charge passes every second. 'That's a useful metaphor, which I can grasp,' interrupted Mary, 'but I am always baffled when anyone mentions the word "charge". What exactly is it? It's what puts me off electricity – whenever the word comes up I just lose the plot.'

That question is as deep as it is simple, and we'll tackle it further on. For now, let's just get a picture in our heads of electricity as something that flows. The rate at which it flows is the current, and that's measured in amps. The tiny current in our mobile phones is just a few thousandths of an amp; the massive current passing from pylon to pylon through the overhead cables might run to hundreds of amps. Just one hundredth of an amp passing through your body will give you enough of a shock to make your muscles contract.

'OK', we might say, as Julie did cautiously on one occasion, fearing a torrent of technical terms, 'why do you need volts then?' This fundamental question brings us to the next important idea about electric current – its power. It is this that really matters for most practical purposes, not just the rate at which it is passing by. A torrent cascading down a mountain and a leisurely river meandering across a plain may each be moving a thousand litres of water every second, but the torrent would have greater strength. This is the reason why early water mills were usually located in hilly regions. They powered the machines of the early industrial revolution. The greatest power in a river comes from fast flow combined with a sharp drop in height. We might have noticed this at home. Sometime the taps higher up a house are less powerful than those at the bottom because there's less of a drop from the water tank in the roof space. Bathroom showers can sometimes be rather weak because there's not much of a drop from the tank that is supplying them.

Similar considerations apply for the power of electricity, which depends both on the flow and the 'drop'. Current is the flow and voltage is the 'drop'. Imagine a water mill: the drop in height from the feeder stream to the wheel is like the voltage. Increase the drop and you increase the power of the turning wheel. The power of the wheel would also become greater if the rate at which the water was flowing (the current) were to increase. The AAA battery used in many mobile phones is rated at 1.5 volts, which means there is a 'drop' of 1.5 between its two ends. When we connect the battery to our mobile phone, a drop of 1.5 volts is created across the phone. If it uses two batteries, a drop of three volts is created.

So much for batteries and low power devices. Significantly more power is needed for our dishwashers or vacuum cleaners because they have motors to drive. For this, the electricity that comes through the mains – the sockets in the wall – is needed. This comes at a much higher voltage – like having the water tank much higher above our shower. We'd get more water flowing per second and it would hit us more powerfully. In the UK it is agreed that the standard voltage of the mains is 230 volts. This latter point struck the discussion group as bizarre. 'Are you saying that voltage is matter of choice? Is it up to governments to decide?' asked Malcolm. Dominic, a former international journalist, recollected: 'I thought it was 240 in the UK and 110 or something in the USA'. 'Yes', added Sonya, on a more practical note. 'I have never been sure whether my hairdryer will work if I take it to France.'

It's true that fixing the level of the mains voltage is ultimately a political decision. In the UK it used to be 240 volts and now is 230 volts as a result of EU harmonisation, while in the USA it is 120 volts. It's not really the voltage that matters for practical purposes, however; it's the power the electricity delivers that counts. As we have already concluded, the power is a combination of the flow and the drop – more precisely, the current and the voltage. So, to get enough power for your house, electricity could come at a higher voltage and lower current or the reverse – it's a matter of choice. The UK system has a higher voltage so the current is lower. Once a country has adopted a standard it must stick to it, however, as all electrical equipment has to be manufactured to that standard.

What matters from a practical point of view is the power rating of an electrical device, rather than the current or voltage. The electric motor in your washing machine or vacuum cleaner needs a certain amount of power to work properly. The power is measured in watts. A light bulb may operate at 60 watts (60W) for a medium light or 100W (for a brighter one), for instance, while your washing machine may require 500 watts of power from the mains. If you want to try some simple maths, the power is simply the voltage multiplied by the current. If your mains electricity at 230 volts lights up a 100 watt light bulb, the current must be 100 divided by 230. In other words, 0.43 amps.

Static

Another common experience of electricity is static electricity. This type can give us a bit of shock when we point our fingers at an object such as a door handle in a dry room. 'Yes, I'd heard that in hospitals they have chains running from trolleys to the floor to prevent static electricity building up. Is that right?' queried Mary. 'I'd heard that it makes the electricity run down to the Earth, by-passing the rubber wheels,' Malcolm recalled. We may notice static electricity in other places; for instance, people with fine hair may find that it doesn't settle down when they comb it because of the static charge repelling the hairs from each other. 'Isn't that why you sometimes get a crackling noise when you take off a nylon jumper in a dry room? It's static electricity causing little sparks, isn't it?' Sarah added. The idea of static electricity leads us neatly into the next big question: what exactly is it that does the flowing when electricity flows? 'Is it something physical? Is it charges or atoms or something abstract like energy?' as Michelle put it. Nor should we forget Mary's earlier question, which we haven't yet answered: what exactly is charge, anyway?

The properties of static electricity were investigated and analysed long before people understood what charge was. This fact came as a bit of a surprise to the group. 'How can you investigate something if you don't know what it is?' Rosie asked. On reflection, however, it becomes clear that this is very much what science is like: we play around with things, try out experiments, explore the unknown, well before we understand what is actually going on – that, indeed, is why we do it. In practice it's full of uncertainties and partial understanding; very different from the impression of science as rigid, all-knowing and rule-bound that is so easily picked up. A classic example was Gregor Mendel's research on peas in the mid-nineteenth century, which conceptualised the idea of a heritable factor long before people knew about genes or DNA.

Originally the most widely accepted idea was that electricity was a kind of fluid in metals. As with the even earlier theory of 'caloric', a supposed fluid that explained the flow of heat, this intuitive early model of electricity had to be abandoned when experiments disproved it. What we now know is that *all* substances are made of electrical constituents. More precisely, all substances are composed of atoms (under normal conditions) and the interior parts of atoms are electrically charged. The modern concept of an atom, developed in the early twentieth century, is of a tiny, hard nut of positively charged particles (called protons) sitting at the centre (called the nucleus), with an equal number of negatively charged particles (called electrons) existing around the nucleus. These negative electrons are spaced apart, far from the centre, and are normally held in place by attraction to the positive nucleus. Overall, atoms and the substances that are made up from them are perfectly neutral;

that's because the amount of positive and negative charge in them is exactly equal – it balances out. A good thing too. It's why ordinary things such as tables and chairs, people and planets are just there, sitting around in uncharged neutrality, rather than rushing around in a frenzy of mutual attraction.

However, in special circumstances some otherwise neutral materials can become charged. The plastic of a comb is one of these, and so is the polythene of supermarket bags. Have you ever been irritated trying to open up a carrier bag or bin liner, when both sides seem to stick firmly to each other? Some electrons from the atoms on the outer surface of these materials can be pulled off, leaving the material electrically unbalanced. A charged-up sheet of plastic can then attract a neighbouring piece. You can check this with a comb. Run it through your hair to charge it up, then bring it close to some small scraps of paper or hairs or dust. They are drawn to it.

So electric charge is carried by particles – usually electrons in the situations we are talking about now. It's these electrons that flow in a wire or any piece of metal. A distinguishing feature of metals is that they have electrons in them that are permanently detached from the atoms they came from and are free to move around. They are buzzing around inside pieces of metal all the time, in random directions, getting nowhere. But, if we connect the two ends of a battery across a piece of metal, the voltage drop that we have applied makes the electrons everywhere in the metal drift towards the positive end of the battery. This flow is the electric current. It's caused by the voltage drop that the battery provides. An AAA battery gives a 1.5 volt drop and causes a certain level of current to flow; two batteries would give three volts and would double the level of current. The same goes for electricity from the mains, but in that case we are applying a much larger voltage drop of 230 volts.

'OK, but you still haven't told us what you mean by charge,' Mary repeated patiently. She had noticed that plenty of people use the word freely, but when she had asked a friend what it actually meant he had no real idea – he had just picked up the word and got used to using it with his car battery and mobile phone. It's not just to be awkward that I've delayed answering the question: it's to avoid having to introduce too many concepts at a time. It's easy to get confused or forget something mentioned earlier and then give up. Remember learning to drive: gear, clutch, brake, accelerator, steering wheel, indicators, mirror? To recap: the main idea we have now put in place is that electricity flows because charged particles (electrons in the case of metals) flow freely through conducting materials such as the copper in a wire, when a voltage drop is applied across it.

Now we can tackle Mary's question: it's a fundamental one and, to some extent, philosophical. Charge is a property that we have invented to explain what we observe; it's not a visible thing, such as colour, nor anything tangible, such as stickiness. It's a concept that was developed in the eighteenth century to explain what happens when electrical substances are brought close together. Etymologically the word arose to imply 'filling up with electricity', in the same way as we might charge a cannon. The original substance was the precious stone amber (the Greek name for which, *elektron*, was adopted to name the particle). When a piece is rubbed with a cloth it becomes electrically charged. We can try the same sort of experiments today with bits of plastics or, better still, pieces of polystyrene foam used in packaging.

This observation alone shows there must be some kind of force acting between materials in this situation. But how can we account for the fact that this electrical force sometimes attracts and sometimes repels? After all, not every force is like this: a falling stone is always attracted towards the Earth, never repelled. So, although electricity was not well understood at the time, scientists came up with an ingenious theory to explain all this. They imagined the cause of the electrical force comes in two flavours; when they are both the same we get repulsion, when they are opposite we get attraction. They could have called these two flavours anything – 'blue and red' or 'Fred and Ginger' – but the names finally assigned were 'positive and negative'. These names indicated that the 'flavours' were opposite to one another, in the sense that equal amounts of positive and negative charge cancel one another out.

So charge is an abstract concept used to explain electrical force: the greater the force, the greater the charge that is causing it. Charges of the same type repel each other; unlike charges attract each other. There just isn't any more tangible idea of what charge is; it's a fundamental property of the particles of which matter is made that helps to explain electrical forces. I suppose the same applies more widely, for many concepts we use every day. What actually *is* gravity, for example? These are all abstract concepts introduced to make sense of what we observe.

Electronics

There's another aspect of electricity that we've all become aware of in recent times, even though its meaning is not widely understood. Sarah captured this when she asked, 'What is the difference between electricity and electronics? Sometimes we're talking about gigantic cables and pylons that can kill in an instant; then the next minute we're talking about some tiny battery in a hearing aid that wouldn't hurt a fly.' It's true, there are two quite different worlds from the everyday point of view: the world of 'big' electricity, where it's the power that matters, and the micro world of electronics, where it's information and control that count. Electricity macro sense began to be explored systematically in the eighteenth century, but the world of electronics only opened up in the early twentieth century after the discovery of the electron. Fortunately from a scientific point of view (and an educational one), exactly the same concepts are involved, and the same terminology and units as well.

Put simply, to light up our living rooms, heat up our irons or turn the motor in our vacuum cleaners we need quite a bit of power. One bulb needs 60 or 100 watts; so a few lights and a machine or two switched on means you typically need thousands of watts of power (called kilowatts, kW) to supply your needs at home. But to power up a mobile phone takes less than a single watt. The main job of an electronic device such as a computer is to send tiny bursts of current round circuits to represent numbers and codes for instructions. These minute bursts are controlled using transistors that are embedded in tiny pieces of silicon – the so-called 'chips'. The small amount of power needed in these devices is mainly used to light up a screen, or to activate a loudspeaker or cooling fan.

'Can we pause for breath here?' suggested Mary. 'We've covered a lot of ground; can we recap? Am I right: you're saying that it's the volts that drive electricity round a circuit and the amps that tells you how much is flowing past at any given time?' She was right: that's the broad picture. Don't forget that it's the power that counts for most purposes in the home, and we get that by multiplying the amps and the volts. For example, if the power rating of your cooker is 2000 watts and the voltage of mains electricity is 230 volts, the current must be 2000/230, i.e. 8.7 amps. This is useful when we need to work out which fuse to buy for a plug. 'Hang on, hang on a minute,' cried Julie, vigilant as ever for any new concepts slipped in unannounced. 'What's a fuse? What does it do? They lie around the house and occasionally one needs replacing. What's going on there?'

Good question: it illustrates another important feature of electricity, as well as being of practical importance. The original meaning of the word fuse is to melt. In electricity it's used to protect valuable equipment that could get burned out accidentally and to prevent electrical fires in the cabling. If, by misfortune, an excessively large electric current were to pass through a device, say a washing machine, it could heat up the wires so much that they melt, destroying the interior workings. This time it was Sonya's turn to raise a follow-up query. 'Why should the wires heat up when a current flows through?' Another fundamental issue had been opened up.

Resistance and heating

The answer is that, to some extent, everything that carries an electric current also offers resistance to it. It's rather like a cyclist travelling along a road or a parachutist falling through the air: there's friction or air resistance opposing them all the way. You have to overcome that, or at least balance it, to make progress. In a similar way there's resistance to the flow of electricity. You may have noticed that overcoming resistance often has the effect of heating things up; think of when we rub our hands together or sandpaper some wood. It's the same with electricity passing through a wire or any other conductor (apart from the very special case of 'superconductors'). As the electrons move through the wire, they collide with the atoms of the metal – past which they have to travel, rather like a ball passing through a pinball machine. Some of the energy of the moving electrons is transferred to the atoms of the metal, causing them to jiggle about more – it's this movement of the atoms that we experience as heat. We may have noticed this heating effect with our computer or mobile phone.

Getting back to our washing machine (or any other electrical device), the wires inside will offer resistance as a current flows around, but under normal conditions the small amount of heat this generates will easily dissipate. But if by some accident two bare wires were to touch, the current might suddenly surge, causing them to heat up sufficiently actually to melt. This could wreck the motor or even the entire machine. The job of a fuse is to melt *first* in any accidental situation where the current gets too high, before the precious machine does. By melting before the wires in the electric motor do, the fuse breaks the circuit, stops the current flowing and saves the washing machine.

We therefore need to choose a fuse that allows enough current to our washing machine, but not too much more. In the example above, the current was 8.69 amps, so we need to choose a fuse that allows this amount of current to pass, that is, a 13 amp fuse rather than a 3 amp one. These are the type of fuses that you buy in a shop and insert in the plug attached to an appliance. If a piece of electrical equipment fails to work, it may be simply that the fuse in its plug has blown. If this is the case you can test the fuse by seeing if it works in the plug of a different device, such as a table lamp. There are also fuses that protect your mains or lighting circuits as a whole. These are the ones that can plunge your whole house into darkness if they blow. Nowadays many homes don't have a central fuse box. Instead a box full of 'trip switches' operates to cut off the current if it gets too high by accident. This saves us having to replace a fuse.

The mains

'I've got a simple question,' interjected Sarah at this point. 'I don't know if it is relevant, but what are the three holes in an electric plug all about?' Once again, an apparently simple question opens up an important scientific issue: where mains electricity comes from and how it gets to us. It also raises a point about the different ways in which electricity can move. Up until now we have talked about electricity moving in a circuit, meaning that the electrons flow round from point to point in an endless loop. On the way they pass through the battery that energises them and the various switches and devices in the circuit, such as a light bulb or a screen or a motor. They activate the device as they pass through – light up a bulb, spin a motor or whatever. This is how a torch or mobile phone works, for example, with the electrons driven by the voltage drop between the two ends of the battery.

But mains electricity doesn't move in the same way. There's no battery. Instead a large and powerful generator in a distant power station spins around at great speed producing a voltage. This voltage is not steady like a battery; instead it rises and falls regularly. A bunch of wires spinning in a generator in the power station rushes past a giant magnet 50 times a second. It is the interaction between these wires and the magnet that generates the voltage. The voltage not only rises and falls as the generator spins, but actually reverses every half turn. This alternating voltage gives rise to an alternating current which is depicted in the graph (Fig. 16.1).

As the graph goes from positive (above the line) to negative, the current reverses direction. One minute it's flowing to our homes, the next minute it's flowing back the opposite way.

'I'm not sure if I get what you're saying, but it does sound pretty ridiculous,' intervened Julie, keen to grasp this bizarre world of

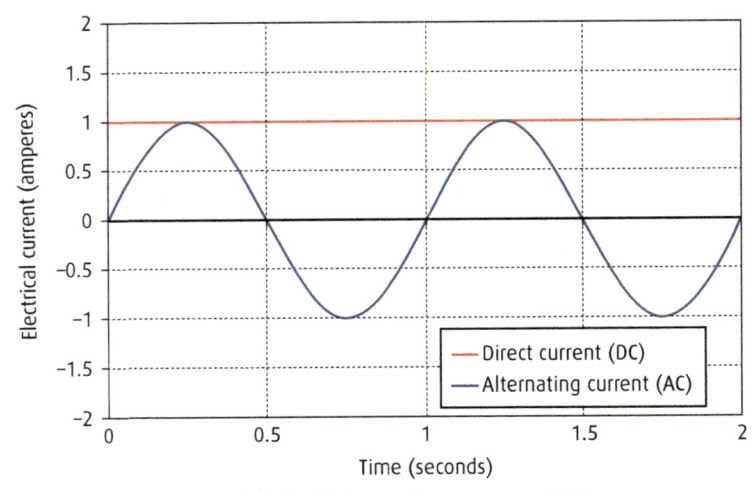

Fig. 16.1 The rise and fall of alternating current (AC)

alternating current. 'It sounds as though the power station is sending out a load of electricity and then you're just sending it back again, all in a fiftieth of second – what's the point?' It's true that this is what is happening to the electrons in the wires that run between the power station and our homes – they really are shuffling back and forth 50 times each second. But what such a description misses out is what the electrons in the wire are doing in this rapid hokey-cokey. If they are passing through a heater, for example, they are heating it up as they pass back and forth, just as much as if they were passing through continuously in one direction only. Think back to the analogy of sandpapering a piece of wood. As the movement goes back and forth, heat is generated with each stroke, regardless of the direction. The same goes for anything else the current may be passing through: a light bulb, a motor or a computer screen, for example.

'Wait, wait, wait, Andrew,' Julie cried out, expressing understandable disbelief. 'Are you really saying these blessed electrons have to travel maybe hundreds of miles from the power station to your house 50 times a second – it's crazy!' I remember this foxing me when I was first told about alternating current at school. Later I was shocked to discover that this is not at all what happens. In turns out that electrons merely drift along a wire at extremely slow speeds – less than 1 mm per second typically; they don't rush by as you might reasonably expect. 'So how do they get all the way from the power station then, if they're so slow?' persisted Julie. The answer is that they don't. What I had forgotten when

I first encountered this strange fact was that the wire is absolutely chock full of electrons all along its length, not just at the start in the power station.

Imagine an oil pipe completely filled with oil. As soon as we push some oil in one end, some more oil pops out the other end. It just moves a bit all along its length. So when the power station creates a voltage drop along the wire, all the electrons move together one way, then they all shuffle back the other way, 50 times a second. They don't get far – a tiny fraction of a millimetre – but, just like sandpaper rubbing back and forth, they still have their effect on your light bulb or washing machine motor.

'To get back to my question: what's all this got to do with the three holes in the plug in my kitchen wall?' insisted Sonya. First of all, a slight correction: it's the socket in the wall, not the plug; the plug is what you put into it. More importantly, we now can understand what the so-called 'live wire' is (the brown one in the UK). This is the wire that connects your socket directly to the power station, through all the cables under the street and up in the air. This single wire connects the voltage that's rising and falling 50 times a second at the power station to your socket. It's live, it averages 230 volts and it's dangerous – not to be touched. But for the electrons to rush backwards and forwards through your washing machine or bedside lamp 50 times a second there has to be a reservoir for them to empty into temporarily, 50 times a second. The entire earth acts as this reservoir. So, as the voltage from the power station rises and falls, it pushes the electrons back and forth at every point along all the wires from the power station to your washing machine and on into the earth. So the earth acts as a gigantic pool, taking in electrons then sending them back, 50 times a second.

To make this happen, there is a second wire in the socket, called neutral. This has been connected to the earth, literally, somewhere around your home. 'You mean there is a wire actually stuck into the ground?' Yes, seems strange, doesn't it? It may be some kind of a spike that is driven into the earth, with a wire attached to it. Or the wire might be attached to a metal pipe used to bring gas into our homes that goes underground.

Reverting to the original question, we now see that one hole in the socket brings the wire from the power station (the 'live' wire), while another is connected to the earth locally (the 'neutral' wire). These two are the lower two holes in the socket. 'OK, so what's the third one for – you don't always have three abroad, do you?' Sonya asked. It's true, not all countries use three wires. In the UK the third hole in the socket links to another wire that ends up in the earth, but it has a completely different

Fig. 16.2 Wire connected to the earth

function to the neutral one. It's a safety extra. When a radio, hairdryer or kettle is made, any outer metal surfaces that we might touch are connected to the third prong on the plug, called earth (or ground in some countries). This means that, if something went wrong and the outer surface of the device accidentally became live, it would just connect straight to earth (Fig. 16.2). This way, if we were to touch it, the current would barely pass through us, going along the easy route direct to earth instead. We would be spared a dangerous shock. The third hole is thus only used to protect us, in the very unlikely case of a device accidentally becoming live.

Conclusion

This foray into mains electricity – the sockets in our homes, the power station, the flow of alternating current – has introduced a plethora of concepts about the basic nature of electricity: electrons moving along a wire, the resistance they encounter, the generators that create the voltage to push them, and more. It proved more than enough for one discussion session, but still left many more questions hanging in the air. Mary wanted to know what all this has got to do with the way our nerves work. 'They're electrical too, aren't they? Are they like wires linked up to batteries?' 'How do batteries work, anyway?' interjected Sarah. 'They're completely different to mains, surely. After all, they run down after a while.' 'Car batteries are filled with acid, aren't they? Why's that?' asked Michelle. 'What about those birds you see perched happily up on overhead cables – surely there must be thousands of volts across their bodies?' And so the questions continued to flow. We'll have to leave these and others ('What are electrolytes?' Sonya queried, in a last ditch attempt) to another occasion. There's plenty more to talk about, but at least we've covered some of the basic concepts in this chapter. They should help you make more sense of some of the electrical mysteries you encounter in everyday life.

Electricity, like many aspects of energy, crops up as a topic in almost all areas of science. It is a central topic in physics, but is also key to understanding reactions in chemistry and the nervous system in biology. This is typical of many of the most fundamental concepts in science. The work of a professional scientist regularly crosses the boundaries of school subjects. A chemist may seek out a particular species of plant in search of chemicals for a new medicinal treatment. In so doing, she would find herself criss-crossing between botany, chemistry and pharmacology, quite possibly engaging with mathematics and computing in the process. Real life problems require multi-disciplinary responses.

What is true for the practising scientist also holds for us as ordinary citizens. Whether we are worried about a child's illness or checking out a faulty heating system, we are more concerned about illuminating the problem than about confining ourselves to the subject matter of biology or physics. For this reason, conversations in a science discussion group that start from everyday concerns tend to range freely over the disciplines. What is more, they are just as likely to delve into the nature of behaviour and emotion as the structure of molecules or behaviour of pulleys. How we recognise faces, why we take risks, what makes us love our children are typical examples of issues that have arisen in discussion groups. With this in mind, the next chapter draws on a number of discussions that began with questions about behaviour. The intrinsic fascination of the topic led one group to arrange visits to a neuroscientist working in the field. She had been using the technique of functional Magnetic Resonance Imaging (fMRI) to investigate aspects of brain activity related to behaviour and emotion. We explore what emerged in the next chapter.

17
MRI and the Brain

Are the effects of puberty the same in all cultures? Does hypnosis really work? What's happening in the brain when people become addicted? Where's the part of the brain that knows what to do? These questions and many others like them have cropped up in discussion groups over the years. Human behaviour is a constant source of inquiry for people reflecting on their experience of everyday life. It has been a rich area for sociologists studying how we interact in social settings and for psychologists conducting experiments, often at the individual level. The biological processes affecting behaviour are also being researched through, for example, studying the effects of hormones, drugs and other body chemicals.

Advances in brain science in recent years have dramatically increased our understanding of behaviour, particularly with the advent of a radical new technology: the MRI scanner. Familiar now to an increasing number of people, thanks to its amazing capacity to depict the inner parts of the body, the scanner has become a key diagnostic instrument in hospitals. Its role in researching the brain, however, remains something of mystery for most of us. For this reason one discussion group invited a cognitive neuroscientist, Dr Iroise Dumontheil, to join them to see whether MRI research would shed light on some of their questions. Before we see how that discussion went, we can usefully remind ourselves of what was said about MRI scanning itself in chapter 7.

MRI mechanism

Magnetic resonance is a physical technique that has been developing in physics and chemistry labs throughout the past half century. Originally a research tool used to study the structure of molecules, its method depends on the magnetic properties of atoms. Deep inside every atom is a positively charged nucleus which, in some cases, is slightly magnetic. When a strong external magnet is applied to any object it affects the atoms of which it is made, tending to line them up magnetically. Then, if a further pulse of radio waves is applied, the precise position of the atoms can be identified. In medical applications the hydrogen atoms found in water are used for magnetic resonance imaging. The scanning technique enables the locations of the zillions of water molecules in your body to be revealed. The crucial point is that, fortunately, water in different tissues is affected in slightly different ways. This enables the various tissues in your body to be distinguished. Tumour cells, for example, respond differently to normal tissue and so stand out.

A special approach, known as functional MRI (fMRI), is used for studying the brain. It so happens that, when the brain cells in a small area are active, they require an increased supply of blood to give them energy. By good fortune the MRI signal from blood as it arrives, rich in oxygen, differs from that of departing blood, which carries less oxygen. This allows the MRI scanner to detect where there is increased blood flow and hence, indirectly, where brain cells, or neurons, are most active. As a result we can see different parts of the brain 'light up' in the scanner wherever there is intense neuron activity. The scanner can detect differences down to a resolution of about 1 cubic millimetre, yielding a wonderfully precise image compared to previous methods. Nevertheless, even this tiny amount of brain contains several million cells, so the image from a scanner still grosses up a very large number of neurons. It can't tell us what may be happening to individual neurons.

The landscape of the brain

Before we go any further into brain scanning, we ought to look back at the basic geography of the brain outlined in chapter 7. As Mary once observed in discussion: 'We've got the idea that the brain consists of zillions and zillions of individual cells, but how does it work as a whole?' 'You see images of it in books, and in France you might see an actual brain in a butcher's shop, but how do you know which parts of the brain do what?' added Sonya.

Although the organ looks much the same for each of us, it is clear that our brains are not all identical. To take one extreme, a child was once found, through scanning, to have only half a brain, but he was otherwise fine. Not only does each brain look different, they also change over time. Physical changes in the brains of London taxi drivers have been detected after they have spent years memorising the street plan, for example. A study of London taxi drivers, before and after a tough training programme, showed growth in the hippocampus, an area of the brain associated with memory and spatial navigation. Nevertheless the brain has been found to have distinct regions, common to all, and these have been gradually mapped over the past century or so. In the early days, much of this was done by neurosurgeons whose patients had suffered injuries or lesions in particular parts of the brain.

Particular areas of the brain are associated with particular functions such as seeing, hearing, speaking, driving muscles (motor) and responding to sensations (sensory). These are not entirely discrete, however, such as the parts of a motor car engine are, each of which has a unique function. The brain seems to be a more complex system, with some functions linked to several areas and most areas serving many functions. The major, and most familiar, part of the brain, known as the cerebrum, consists of two hemispheres. The outer layer of these hemispheres is known as the cerebral cortex and each is divided into four 'lobes'. Sitting beneath these are a number of other discrete parts of the brain (Fig. 17.1). The latter have evolved from more primitive forerunners of today's human brain and tend to be associated with more basic functioning. The cerebellum, for example, is associated with

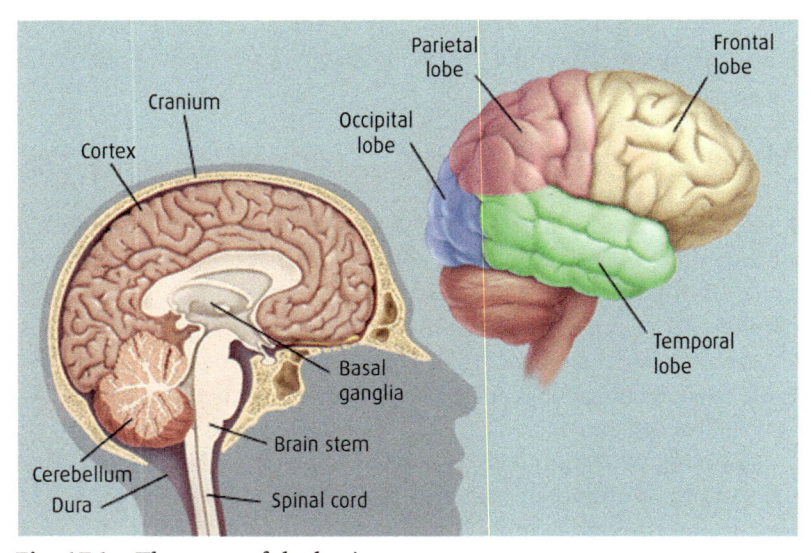

Fig. 17.1 The parts of the brain

fine-tuning our movements and the brain stem regulates our heartbeats and sleep cycle.

The diagram also shows the positions and names of the four lobes of the cortex. These are often referred to when neuroscientists describe their discoveries, similar to the way we refer to the different continents when describing events around the world.

'We know there are two halves to the brain,' Sarah affirmed, 'but what about the left and right brain that people talk about? I've heard that the left side is more logical and the right side more creative.' There are indeed two similar-looking halves to the brain, left and right, and there is a popular idea that each is associated with particular qualities. This is an active area of research, however, and new evidence seems to be suggesting that it's not that simple. Brain processes generally seem to draw on both halves and the brain often seems able to compensate when there is any injury or damage to one area by using the unimpaired half. Even more surprising is the variation between individuals. For example, for 95 per cent of right-handed people the left half dominates for language, whereas for left-handed people the right side is dominant for only 19 per cent.

Life stages

With the basics of fMRI scanning and the regions of the brain clarified, the members of the discussion group were ready to put questions about the issues that fascinated them to Iroise, the cognitive neuroscientist. The first was about how ageing affects the brain. Older people today are encouraged to remain mentally active: to tackle crosswords, play Sudoku and decipher flat-pack furniture instructions. 'Does this mean you can develop your brain as you grow older?' Julie asked. 'Or do you just get a fixed number of brain cells at birth and that's it?' 'I think you can get changes as you go along,' Sarah observed. 'You hear about people who have suffered a stroke and then gradually recover their capacities.' 'Use it or lose it,' added Michelle succinctly.

Broadly these perceptions about the ability of the adult brain to renew itself are correct. There is increasing evidence that the adult brain goes on developing throughout life. It is no longer thought that you are given your full brain power at birth. As Iroise clarified, we do have the maximum *number* of neurons when we are born, but it's not just the number that counts: it's the number of *connections between them* (known as synapses) that really matters and this is relatively small at first. As an infant develops, however, each cell develops branches in a process called 'arborisation' (after the Latin for 'tree') and these make connections with other cells. The density of connections rises and reaches a peak between one and two years of age; thereafter it declines. The remarkable fact is that those connections that get used tend to be maintained, but those that are not used will disappear – a process colourfully known as 'pruning'.

This important discovery means that, as infants get exposed to stimuli, the corresponding brain connections get firmed up. Where there is no such exposure, however, they wither away. Rosie recalled a television documentary about orphan children born in Romania who had suffered extreme isolation in institutions and whose mental faculties had suffered as a result. This reminded Jean of experiments she had heard about in which some young animals had been brought up in a stimulating environment and others in a deprived one. As we might expect, the latter failed to develop as strongly as the former.

'This process for developing the brain seems a little odd,' Mary commented. 'Why are all these connections made in the first place if they are just going to be pruned away?' The explanation seems to be that each connection consumes energy, so the number needs to be kept to the minimum necessary. Also, having too many connections simply adds to the overall 'noise', obscuring the all-important signal. 'But what if some are pruned that are needed later?' worried Emma. Research shows that much depends on when the pruning happens. For example, after a certain age, kittens reared with only vertical lines in their visual environment are unable to see horizontals. Pruning comes earlier for motor and sensory neurons and later for the cognitive ones associated with thinking. In general, pruning is to be seen as an essential aspect of brain development; there are some suggestions that autism may be associated with an inability to prune enough. Rosie was curious about whether the brain can generate new neurons, not just new connections. Apparently this is possible in certain parts of the brain – the hippocampus, for example – but these are few. The creation of new neurons (known as neurogenesis) is an active area of current research, and possible links with mental health conditions, including depression and schizophrenia, are being explored.

Adolescence

The dramatic changes in brain structure that we now know take place during infancy come as something of a surprise for those of us brought up years ago. The old ideas that the number of brain cells was fixed at birth and that brain connectivity increases over time have been knocked for six. It turns out that brain connections are sculpted away, rather like a Michelangelo figure emerging from a block of marble. The number of connections peaks at the age of two, though new ones can be forged later in life. Repetition of mental actions helps with this. 'It sounds a bit like physical exercise, where repetition strengthens your muscles,' Sarah commented. This is clearly of relevance to education and to all kinds of performance activity.

The life stage that our questioners pounced on next was the teenage years. 'Why are adolescent boys so difficult?' asked Mary with a deep sigh, recalling her personal experience as a parent. The central point of discussion was whether the kind of behaviour depicted humorously in Harry Enfield's character Kevin was universal, and whether it was a cultural or biochemical phenomenon. As Sonya put it: 'Is the social effect of puberty different in different cultures?'

Our neuroscientist confirmed that some aspects of behaviour during puberty do indeed differ in different societies, though there are also many general effects. For example, it appears that increased risk-taking behaviour in adolescence may be a consequence of different regions of the brain developing at different rates. Those linked with inhibition and control lag behind those linked with pleasure and reward. A 2008 review for the National Institutes of Health in the USA proposes a neurobiological model in which heightened responsiveness to rewards and immaturity in behavioural control areas may encourage adolescents to seek immediate rather than long-term gains, perhaps explaining their increase in risky decision-making and impulsive behaviours. Of course biology is not destiny: patterns of behaviour vary considerably from one individual to the next. During puberty hormones also affect emotional reactions, but have less influence on the control function. 'It seems like a battle between the hormones and the executive function,' Julie reflected. 'Is there an evolutionary reason why the teenage brain is like this?' asked Jean, seeking a deeper explanation. Our neuroscientist suggested that there could be. After all, at that stage in life, on the brink of adulthood, early humans would have needed to take risks to survive: to find a home and secure a mate.

Rewards and risks

The special research interest of our neuroscientist was the teenage brain. In one branch of her work Iroise was looking at the effect of rewards, involving trials in her research to see if reward incentives improve the brain's capacity to concentrate and exercise control. Rosie was curious about how the brain could physically represent opposed tendencies. 'Does deferred gratification occur in the same part of the brain as immediate gratification?' she asked. The answer was clear: no, the two are quite separate. Deferring your pleasure requires you to think, and that involves the frontal cortex.

The choices we make and how we behave are not, of course, a purely individual affair; social aspects are equally important. Thinking about other people involves activation of both the social and the thinking parts of the brain. Experiments have been devised to test this idea. Studies of the effect of an audience suggest that people's performance can improve (or in some cases get worse) when they are being watched by other people. In the scanner, simply being told that one is being watched elicits activation in the social brain.

Iroise mentioned a famous study which investigated social influences on the teenage brain. The study involved a car driving simulation which included risky road junctions where traffic is crossing. The number of accidents that occur in the simulation game is a measure of the risk participants take (for instance, jumping a red light to complete the circuit faster). When they are on their own, adolescents and adults ended up with the same number of accidents, but in the company of friends the number of accidents experienced by adolescents more than doubled. For adolescents the social context seems to have a significant impact, encouraging them to behave differently.

Rosie wanted to probe beyond the average to the individual level. 'Do individuals vary in their risk/reward reactions?' she asked. It looks as though they do. Genetic differences between individuals can be observed, for example, in the dopamine system. Dopamine is a naturally occurring chemical carrying signals from one cell to the next in the brain (a neurotransmitter). Dopamine can be deactivated by an enzyme (known as COMT) and the gene that produces this enzyme has been found to vary in some individuals, affecting their dopamine activity. It is possible this genetic difference may be associated with mental conditions, including schizophrenia and bipolar disorder. At the time of writing research is inconclusive and it remains a contested area.

This level of discussion about the link between chemicals in the body and brain scans prompted Julie to make an important point about scientific methods. 'There must be an overlap between brain chemistry research and brain scanning with MRI,' she observed. Her point emphasised the fact that scientific understanding often emerges as a result of multiple studies using quite different methods. The two approaches are complementary: fMRI scanning gives low definition information over clusters of thousands of neurons, whereas some brain chemistry experiments can reveal more fine-grained information, for example at the level of a single synapse.

Learning

Throughout most of my life, the workings of the brain have been an obscure area of biology, even for the excessively curious. Little has been known about it among scientists, and even less by the lay person. By contrast the explosion of research in recent decades, fuelled by powerful new instruments such as the MRI scanner, has opened up the field dramatically. As we learn more about it, the enormity of brain chemistry's influence becomes ever clearer. This rising awareness is reflected in the variety of questions on neuroscience that arise in discussion groups. How is learning represented in the brain? What about memory? What's happening when we deceive others? What about addiction, mental illness, criminality and so on? To this panoply of questions our neuroscientist visitor responded patiently, topic by topic, starting with the issue of learning.

Learning has been shown to lead to measurable changes in the brain. In particular, it is thought to change connections between neurons. Learning to juggle, for example, with its endless repetitive practice, leads to an increase in the density of the cortex, which is thought to reflect a local increase in the number of synapses (the places where neurons connect to one another). As the procedure becomes routine, this growth in connectedness settles down and ultimately returns to its previous level, even though the newly acquired skill remains undiminished. During the period of learning, the increase in connectivity is only temporary. An interesting observation of practical importance, from research on skill acquisition, is the correlation between those children who are the first to learn to skip and those who are the first to learn to read.

Patrick, himself something of an amateur musician, asked if this was in any way like the so-called 'muscle memory' that enables musicians to play very complex music almost without thinking – as though on automatic pilot. It appears that it is. When a new skill is being acquired the prefrontal cortex (the thinking part) is initially used. As the skill gradually becomes more automatic, however, the posterior parts of the brain become activated and prefrontal (thinking) activity diminishes.

Celia has an admirable habit of listening quietly in discussions for some time, then framing an interesting and original question. Confronting the fundamental mystery that many of us feel still surrounds the brain and its workings, she asked, 'Where is the part that knows what to do?' This deeply philosophical question led, unexpectedly, into discussion of aspects of brain functioning of which we are not conscious at all; they are simply automatic. There are vital skills we perform without being aware of what we do and which cannot be verbalised. Walking, for example, doesn't work if we consciously try to instruct our limbs. Some of the motor areas involved in physical actions are located on a so-called 'motor strip', which runs from one side of the cortex to the other. Each part of the body is represented at some point along the length of this strip. 'It sounds like a homunculus', as Rosie put it, referring to the ancient notion that a miniature human being existed inside every person. In this area of the brain, specific movements of different parts of the body are initiated. The more profound question of how our consciousness of what we are doing is represented in the brain is a topic of active study, but one that wasn't taken forward in the present discussion. Its time will come, no doubt.

Research on learning suggests that motivation and attention affect the ability to remember. Current research by our neuroscientist on the role of memory in learning confirms what we sense intuitively: that we remember better if we repeat things. Research is currently investigating how long we should expect to wait before trying to retrieve a memory. For teachers, it looks as though it may be best to wait a number of days before asking students to recall their earlier learning.

Inspired by the idea that neuroscience may be beginning to affect teachers' practices, Celia became curious about the impact basic research has on society more generally. 'What happens to your research after you have got some results?' she asked. Iroise responded by outlining the complementary ways in which research on the brain contributes to theory and practice. Results are normally published in academic journals and thereby add to our basic understanding and to the development of new theory. Sometimes, however, results can also be applied to real-life issues in the classroom. In education, for example, brain research shows that, when children are exposed to entirely new and strange concepts, such as the Earth being spherical rather than flat, their previous naive ideas are not always expunged. Instead they may hold on to these alongside their newly gained scientific knowledge. MRI scans in adults show this is represented as conflict in the brain, and is associated with activation of the cingulate cortex. As a result, we may need to start seeing learning not so much as a process of replacing naive ideas, but more as one of inhibiting them. One practical application of such research is the development of a tool that helps children to inhibit their naive interpretations.

Social issues

Of course the brain doesn't only have to deal with practical matters, such as learning to read or responding to hunger; it also has to process social and emotional influences. A question that fascinated Jean was how we are able to recognise different faces so effectively. Apparently a distinctive pattern lights up in an MRI scan when the subject views a face. This is an active and contended area of research, but it appears that several different regions of the brain are involved, some of which respond to specific parts of the face – nose, eyes, etc. – and others of which respond holistically to the entire face. When we see different faces in an unfamiliar ethnic group, we do not at first see much difference between them. Later, with repetition, we begin to discern differences. It's rather like recognising regional accents in a foreign language, as our neuroscientist put it. In a similar way patterns develop for other kinds of stimulus that occur frequently. A keen bird-spotter can distinguish anatomical details that elude other people, and motor fanatics perceive cars in a similar way. Interestingly, babies are able to distinguish the facial features of different monkeys better than adults can.

The group was keen to know more about the way in which the brain represents social issues in general. 'How does the brain handle embarrassment?' Rosie asked. Social emotions in general involve imagining what is in another person's mind. This contrasts with other kinds of emotion, such as the feeling of disgust you have when you see rotting food. MRI scanning is beginning to open up the neuroscience of social emotions. A review of research into this indicates that embarrassment, an emotion evoked by some kind of social transgression, involves activity in part of the brain that represents conceptual social knowledge (called the anterior temporal region). Guilt, on the other hand, tends to arise as a result of a moral transgression. This emotion stimulates a different area of the brain, the prefrontal region, which represents perspective taking and demands for changes in behaviour.

Sonya was interested in how deceit is represented in the brain. It's an intriguing issue: how on earth can the brain hold a pattern that represents knowing something that is true and at the same time represents a corresponding deceit? Does the brain fool itself? It turns out that deceit is indeed studied in neuroscience. Experiments have been conducted using functional MRI with actors lifting up a box. People watching videos of this action judged whether or not the actors were trying to deceive them about the real weight of the box. When they judged the actions as being intended to deceive, parts of the brain associated with social interaction and emotion (the amygdala and anterior cingulate cortex) were activated. The conclusion was that this might suggest that being deceived evokes an emotional form of response.

Mental illness

'How much do we know about the brain and mental illnesses – schizophrenia, for example?' asked Jean, moving on to yet another way in which neuroscience impinges directly on everyday life. Apparently people with schizophrenia have a particular unusual ability: they are able to tickle themselves, something most people simply cannot do. The reason is that the normal brain anticipates the sensation of any action, for example picking up a pencil. Then, when the action is actually happening, the sensation is attenuated. For some schizophrenics, however, this attenuation appears not to happen. Trying to identify any structural deficiencies that are associated with schizophrenia is apparently proving difficult. However, some environmental factors do appear to play a role, including abnormal levels of some neurotransmitters (dopamine and serotonin) and certain kinds of complication arising in childbirth.

'It's odd that such a debilitating condition continues to survive in the population,' commented Mary. 'You'd think evolution might have gradually eliminated it over millions of years.' It's a good point and one that needs explanation. Schizophrenia occurs throughout the world and its incidence is similar everywhere, at about 1 per cent of the population. Since it hasn't been eliminated through evolution, some people think it might be an extreme version of something that actually confers some advantage across the population as a whole. One suggestion is that schizophrenia is a 'disorder of language' and that the illness is an unfortunate consequence of the development of human speech, expression and creativity. It is as though schizophrenia could be a negative outcome for a few, outweighed in evolutionary terms by cognitive and language benefits for the many.

'What about psychopaths?' asked Julie. 'Are they born that way or do they develop the tendency through their experiences in life?' Genetic research into psychopathy suggests that a specific gene might be associated with psychopaths who lack remorse. At the same time psychological studies show that, when shown pictures of faces expressing fear, some psychopaths are simply unable to name the emotion. As Julie was quick to point out, this raises serious moral dilemmas. For example, if it is known that certain brain defects lead to psychopathic action (kleptomania or paedophilia, for example) should the sufferer be punished for their crimes? Or, were it to become clear that depressed people with a deficiency of serotonin feel happier when serotonin is administered, should it be given to them regularly?

These issues arising from the neuroscience of mental health provoked discussion about the relative roles of nature and nurture. Iroise pointed out that some things are learned, some are not, and others lie in between. Scientists draw a clear distinction between a disposition towards a mental health condition and trigger factors that may precipitate it. For example, a psychopath from a more privileged background may not be provoked into violent actions that another equivalent one, from a disadvantaged background, might be.

Conclusion

An evening with a practising neuroscientist proved a truly fascinating experience for members of the discussion group. It provoked questions and observations over a huge range of human affairs: from adolescent development and risk-taking to the processes of learning and deceiving. Questions spring easily to mind because the workings of the brain touch on so many aspects of our lives. Yet for most of us, paradoxically, our everyday understanding of it is very limited. This is hardly surprising given that the big leap forward in neuroscience made possible by MRI scanning has only been made in the last few decades.

Discussion with a neuroscientist managed to convey more than purely factual information. It also offered insight into the nature and methods of contemporary science. A sense of the interdependence of the various sciences emerged as the huge range of relevant methods and disciplines became clear. Not only fMRI scanning, but experimental psychology, pharmacology, anatomy, surgery, biophysics and biochemistry all contribute to an understanding of the brain. Something about the scientific posture also impressed the group: the judicious way in which the scientist answered the broad, everyday questions posed. Answers were always given in relation to quite specific research findings and were usually hedged around with caveats. Findings were related to the specific context of a study and a sceptical approach taken to any attempt to generalise beyond this. This scientific scepticism means that our understandable desire to get complete answers to important questions is often frustrated. 'We don't know the whole story – we have some bits of knowledge from a few specific studies' is frequently the response.

Then there is the issue of bias. Scientists in all fields, including the social and human sciences, are perpetually on the lookout for it; strenuous efforts are made to seek it out and compensate for it before drawing conclusions. In one example, Julie asked whether bias might creep into the recruitment of subjects for fMRI scanning experiments. Our neuroscientist thought there probably was, pointing out, for example, that it's easier to recruit teenagers from public schools than state schools. It's possible that conclusions about adolescent brain activity could be affected by this cultural factor.

In summary, there's a strong sense in the field of neuroscience that tremendous advances lie just around the corner. It's an exciting field with plenty of tools at its disposal and lots to discover. The implications for society and individuals will be profound. Will education be transformed by new insights into the learning process? Will new effective treatments emerge for mental illness? Will we come to see mental illness increasingly as brain disorder? Will our concepts of culpability and punishment have to alter? Whatever the future holds we can be pretty sure this very chapter will be out of date within the decade.

Discussions about neuroscience are particularly engrossing. Some of our deepest existential questions meet with science's equivalent of breaking news. It's no longer a matter of learning about the established verities, such as laws of gravity and rules of refraction. Instead we engage directly with emerging results, fresh from the lab. Much of what is currently mooted about brain function is not finally settled. By exploring neuroscience we enter the real world of contemporary science, engaging with partial knowledge and contradictory findings. Indeed it is inevitable that some of what is currently held to be true will be superseded in years to come as research progresses. Exciting though this acquaintance with science-on-the-move is, it also carries a degree of frustration, as this chapter shows. Important things that we'd like to know about are simply not yet understood; answers to questions of burning importance are often incomplete.

Our interest in brain science is, of course, not purely academic. We want to know if brain science can help us to understand human behaviour: the impulse to learn, the nature of emotion, the path to addiction. In discussion groups, questions about behaviour often probe beyond our own species, *Homo sapiens,* into the wider animal kingdom. To what extent are we humans unique? Do we share some behaviour patterns with our pre-human ancestors? Are language and culture restricted to human beings?

Television programmes about the natural world are immensely popular. Watching different species build their homes, find their mates and manipulate their environment is enthralling. But what drives them? Are they acting out of genetically ordained impulses or making cultural choices? A fascinating discussion once began with questions about loving and aggressive behaviour, and led into the role of evolution in shaping these. In order to pursue these issues in relation to our fellow primates, a visit was organised to meet with an evolutionary anthropologist. The next chapter is based on that discussion. It takes us into the social lives of various species and throws light on what we mean by 'culture'.

18
Animal Culture

Sonya came to a discussion session one evening filled with thoughts about attachment theory. She was studying part time to qualify as a psychotherapist, and had been learning about experiments carried out with baby monkeys brought up with models of mothers in place of real ones. Some of the 'mothers' were made to be warm and cuddly, others cold and hostile. As we might expect, the baby monkey fails to attach to the cold 'mother'. 'It seems to reinforce common sense', as Sonya noted. What's more, research shows the effects of this non-attachment can be long lasting.

Discussion in the group

'I have always been mystified why people get so attached to abstract things like brands,' observed Patrick, choosing a different tack. 'It's the effect of marketing and advertising,' claimed Julie, who had originally studied social sciences before developing a career in IT. 'It's gradually re-engineering society,' she added. 'I heard that it took 10 years for people to accept that the claim by a company that its bras were better fitting than others was just hype.' 'But then there are football clubs and pop idols as well,' added Malcolm, extending the scope beyond fashion. 'They are brands too.' 'Perhaps it's all about tribes,' Julie conjectured. 'But there's a psychological aspect too,' insisted Sonya. 'Men are generally brought up to suppress their emotions, so football is a great opportunity to let it all out. If you look at bereavement, it is a time when adults are allowed to cry – it opens the flood gates. You end up crying about many pent-up things.'

This seemed too simplistic to Marian. 'Italian men seem to express their emotions more directly than others. What do Italian mothers feel about this? I suppose they are part of the attachment process.' There seem to be important differences between cultures in how men are expected to behave. Sonya, who had roots in India, pointed out that the 'heir and spare' issue varies between cultures – the question of who inherits the family fortune and who is left with nothing. 'It depends on inheritance laws,' interjected Julie, recalling the cultural studies that had been part of her university course. 'Law engineers behaviour. Law is a mixture of objective and partisan perspectives.' 'So it's a bit like science, then,' Malcolm commented drily. 'It calls for evidence and puts it to the test. I suppose the difference is that the law always has to come up with a firm decision at the end' he added.

This issue about the role of science struck a chord with the group. Sonya put her finger on a widely held criticism of scientific research – its tendency to prove what the actor John Cleese once described so eloquently as 'the bleedin' obvious'. 'Attachment experiments just seem to confirm what is common sense anyway,' she observed, adding on reflection: 'Science does subject common sense to rigorous testing, though. Sometimes it turns out to be right, but sometimes not. I remember a joke I was told at school about someone who got drunk on gin and water, then on rum and water, then on whisky and water. Common sense told him that obviously water makes you drunk.'

Discussion returned to the influence of culture on behaviour. A topical issue in the news was the revelation that some media celebrities, now in their later years, had been abusing young people regularly since the 1970s. 'It was broadly accepted in that culture at that time, whatever we feel about it 50 years on,' Julie commented. 'Racial abuse was casual then too,' she added. Patrick, who had been reading a book about the history of violence, pointed out that it too had been much more acceptable in the past. Hard though it is to accept, given the atrocities of the twentieth century, research shows that, overall, violence has actually been gradually declining over the centuries – any reading of mediaeval history attests to this.

What is actually happening in the brains of violent extremists? What drives men (it is mainly men) to atrocious acts of war or terrorism? Why were horrific punishments acceptable in medieval Europe or in parts of the world today? Does evolution explain such tendencies towards violence – are they connected to survival over the long term? Does aggression ultimately win out over collaboration? To address these fundamental questions it became clear that expert knowledge was going to be needed from someone involved in researching the field. But understanding the roots of human violence seemed to involve many distinct scientific disciplines. Where should we look: psychology, zoology, evolution, history? Fortunately an anthropologist with expertise in both the behaviour of primates and the role of culture was found in the local university. Professor Volker Sommer proved willing to meet with the group and tackle their questions. What follows is the fascinating sequence of questions and responses from an evening with him.

Dialogue with an anthropologist

Volker's background seemed to confirm the interdisciplinary nature of our subject. He works on evolution, anatomy and physiology and was able to offer plentiful insights into the nature of culture. His central interest is looking at humans as apes, given that both have common ancestors. Volker's research takes him to Africa for part of each year to work with apes; the other part engages him with humans at the university. Working with primates is necessarily a long, sustained kind of research process. It takes many years for a researcher to be accepted by an ape community and for accurate assessments to be made of their varied behaviour. As with human societies, social tensions can build over many years, then break out suddenly on occasions. A snapshot taken over too short a period may miss shifts between peaceful and aggressive behaviour.

'So primates can be peaceful sometimes, then aggressive later?' asked Sonya, confirming that social tensions can alter over the years. 'That must mean they have memories,' she inferred, correctly as it turns out. Apparently a gorilla researcher went back to visit his previous ape community after a gap of 10 years and was hugged by a silverback male gorilla there! Research on memory is carried out through experiments in which the apes are set specific tasks and tested on them; they are then retested many hours later to see whether the tasks had been remembered. On an unexpected sombre note, the group learned that it's quite likely that most wild apes will soon be wiped out by hunting and habitat destruction. Just a few will have a chance to survive, in particular those that are under long-term study by researchers, who have often turned into passionate conservationists.

'So violence in primates seems to be periodic, not a fixed characteristic; it comes and goes. What triggers it?' asked Mary, probing a little further. It's a matter of risk analysis, the anthropologist explained: calculating gains over losses. The main issue is access to resources, such as fruit trees, but also mating partners. In the case of chimpanzees, males patrol the borders of their territory routinely; if they see that a party of neighbours they come across is smaller than their own, they are likely to attack.

Talk of gender differences inspired discussion about pair bonding. It turns out that societies are structured differently in different primate species. There are variations in bonding pattern, just as there are for humans: some are polygynous, some polyandrous, others monogamous, and yet others are polygynandrous (societies in which both males and females have several mates). As our expert summed it up: 'primates are flexible.'

Rosie was keen to explore an even broader issue. 'Do animals have culture?' she asked directly. As you might expect, the expert response was that much depends on how you define 'culture'. To illustrate the ambiguity Volker took the example of how we might eat rice: with chopsticks or forks. At first this might be understood as a cultural difference, but in fact it depends on the type of rice being eaten, and whether this is sticky or not. It's actually an example of the influence of the environment. There are differences in customs such as this between smaller populations within a species, but to qualify as 'cultural' differences need to be arbitrary – of no practical value. For example, some populations of chimps eat ants but not termites; for other populations it's the other way round. There's no nutritional advantage either way – it's purely social and about belonging to a group.

This concept that cultural differences are by definition arbitrary came as something of a surprise to the group. It seems to contradict a common sense notion that some kind of utility lies behind the ways in which we behave. But in fact these differences act as signifiers within animal groups; the key thing is they must be *random,* not dictated by the environment. There are countless examples, as the anthropologist explained, like animals sucking each other's toes as a greeting or splashing around with one another in ponds. 'Sounds a bit like teenagers,' quipped Sonya. 'Quite so,' was the reply, 'they are both signifiers of group behaviour.' The same is true for religions – codes must be arbitrary; as Volker put it: 'You must believe the same nonsense that I believe.'

This idea of religious observance as a set of agreed but arbitrary practices provoked further questions. 'Aren't some things in religion advantageous?' asked Jean, tentatively. Yes, it's true there are practical merits in some cases; it is doubtful that religious rules function only to signify belonging. Jewish eating rules that seem arbitrary today probably reflect the times and ecology at the time they originated.

The taboos around shellfish, for example, may have served to minimise food poisoning.

'I wonder if there are taboos in animal society,' Rosie interjected, developing the cultural theme. 'But then how would you know if there were?' she added as an afterthought. It seems that there are effectively some taboos, such as the group of chimps who don't eat termites, for example. Ostracism can also be observed in primates. Primatologists had seen a female who actually moved from one group to another for this reason. She changed her customs to adapt to the new culture. Primates have personalities; they keep a mental record of the responses they get from others and make alliances accordingly. They know if favours have been returned or not and are quite capable of holding a grudge. Apparently when apes make sexual overtures they even count the number of rebuffs they get and give up after a certain number.

The idea that individual primates may move from one group to another proved something of a revelation and moved the discussion on to considering the individual in relation to society. 'Do individuals join groups?' asked Jean. 'Yes, they do,' was the clear answer. 'Groups split and merge, but individuals must take on the customs of the group.' Behaviour of groups once again put Sonya in mind of human teenagers with their collective fads and efforts to distinguish themselves as individuals. But, as she pointed out, teenagers do at least choose their customs, in contrast to other primates. Volker's view was that this is because humans live in much larger societies, providing scope to move around, meet a greater range of people and have many more choices – and situations to negotiate.

This idea about diversity in groups reminded Malcolm of a book he had read about cohesion in US society. It had suggested that individuals having multiple cross-cutting identities – Pennsylvanian, Dutch, American, for example – helps to maintain social cohesion. The anthropologist suggested this may be a problem for large states: human groups seem to like to fracture. A state needs a narrative that cuts across all groups, and this often seems to work for only so many decades. Primate society is smaller, with fewer layers and smaller sub-groups; the narrative is simpler.

Malcolm's reading of US history also threw some light on individuation. He recalled that in the early years of the colonisation of the land in North America, when land was plentiful, young people were able to move away from the parental home and set up their own farm. Parental authority was therefore less influential, enabling individuals to develop more control over their lives. The ability to make our own choices struck

Volker as better for democracies because it reduces the power of clans – effectively states within a state. In Africa much of the population belongs to one of several large, influential clans and political leaders have to take them into account. He saw the potential for clan development as one of the reasons that some democracies forbid polygamy. Individuation is also reflected in the tendency for religions to differentiate and proliferate. In Africa, for example, hundreds of different churches have arisen as a result of the breakup of traditional religious systems.

After this foray into politics and religion, discussion returned to what drives behaviour in primates: has it evolved or is it learned? The key idea is that there is no such thing as instinct: primates are flexible, adopting strategies according to circumstances. 'But are they conscious of the decisions they make?' asked Julie, keen to see just how close apes are to *Homo sapiens*. This in effect addresses the fundamental question of free will: does it exist or not? For our anthropologist, free will implies there was a cause, and cause–effect relationships are something we human beings tend to impute. We may feel that we are free, but at the same time our bodies drive us to certain ways of acting. Maybe the feeling of free will is something that has been favoured in the evolution of *Homo sapiens*.

These insights into aspects of human behaviour that other primates share inspired a flurry of new questions. What fascinated the group was the underlying question: do our human patterns of behaviour represent choices that we have made or are we simply the product of our evolutionary ancestors? 'Does killing ever happen within a social group of animals?' asked Rosie, no doubt reflecting on the terrible human capacity for murder. Volker confirmed that it does. Grandmother meerkats, for example, are capable of eating their own grandchildren for a reason. They may do this in cases where their older offspring are looking after their younger siblings. If these older ones themselves have babies, the latter may be killed by the grandparent to ensure her own babies continue to be cared for. Whether this should be classed as 'murder' becomes a matter for the individual. Human cultures have developed complex and varied conventions about killing; what might count as murder in peacetime might be defined as heroism in time of war.

Mary wondered whether primate behaviour might throw any light on the forms of religion that early humans may have had. It seems this is mainly a question of the need to identify with the group. The wish to conform arises very early in the development of a species. In one fascinating example, members of one species of macaques were observed eating sweet potatoes on a beach, peppered as they were with gritty sand. On one occasion an individual dragged a potato into the water and found it cleaner and easier to eat as a result. Other macaques subsequently imitated it and adopted a new behaviour. At least most of them did; the older males were reluctant to change (sound familiar?). As our expert conjectured, perhaps given they had less life left they were bearing less risk in continuing with the old ways.

This example encouraged Rosie to ask why cultures are generally conservative about risk-taking. The answer was simply that risks are generally disadvantageous. Adolescence necessarily involves a degree of risk-taking as the young take on the challenges of adulthood. In traditional societies, transition rites to adulthood generally involve some degree of pain for the young person. The loss of ritual markers in modern human societies may explain some of the attraction adolescents have for drugs, body piercing and tattoos. Primates also pass through a juvenile phase; it's necessary for learning. Other animals don't. But after this primates need to make a rapid transition to adulthood – and because it needs to be rapid, adolescence is perilous. Thoughts of adolescents entering adulthood inspired Julie to pose the final question of the flurry: 'Do animals love?' With characteristic objectivity our anthropologist concluded the discussion by replying that 'they have a very similar brain chemistry to humans; they have infatuations and love their young.'

Conclusion

This visit inspired the group not only through its insights into primate behaviour, but also by the quality of discussion. By avoiding technical language or addressing listeners 'from on high', the anthropologist defied the stereotype of the aloof scientist. He talked openly and colourfully about his experiences, but always in response to questions put by the group. Real dialogue took root as questions flowed from natural curiosity; the responses of the expert were direct and to the point. The path of questioning was respected, even as it veered across academic disciplines and jumped from topic to topic. Responding in this way is no easy task for academic experts, accustomed as they are to explaining their subject in their own way.

On this and many other occasions, discussion groups have found scientists only too willing to take time out to talk to them. Perhaps this story will inspire others to gather together, browse the research pages of their local university or research institute and fix up a visit to encounter a scientist in his or her natural surroundings.

Epilogue

If you have managed to get this far, congratulations! Not only have you managed to weave your way through dozens of different real-life contexts, but you've also had a crack at understanding some of science's most fundamental concepts. This epilogue is a brief reflection on the book as a whole and some thoughts about how you might take your interest a step further.

Over the many chapters of this book we've seen the immensely varied aspects of life for which science has something to offer. It may be colour in painting and food, signalling through nerves and hormones or behaviour in adolescents and macaques. Whatever the starting point for inquiry, the discussions have enabled us to delve into almost any aspect of everyday life – and indeed beyond, moving into deep existential issues. Simple observations triggered by curiosity lead easily into encounters with scientific concepts that have been developed and refined over centuries. As a result, we can begin to piece together some of the big ideas that explain natural phenomena, much as we do for the ideas that underpin the social and political world. Just as we make use of concepts of family, government, nation, class and gender to navigate social knowledge, we can start using the idea of molecules, cells, waves and fields to make sense of scientific phenomena.

As we have seen, questions arise easily from many different aspects of our lives, provided we feel secure and confident about expressing them. Sometimes these may arise from a household observation, such as the image in a mirror or cold feet on floor tiles; sometimes from the natural world around us: the tides, sunlight or icebergs. The human body, with its various ailments, is a particularly important source of questions and comments; everyone can draw on personal experience to discuss infection, pain and injury. From such experiences insights into the bodily systems – blood, nerves, muscles, defences – follow naturally. These in themselves lead on to the microscopic world of cells and molecules.

The path of discussion that follows from a question is, however, quite unpredictable. Unconstrained by syllabuses or the structure of academic disciplines, discussion ranges in and out of the separate subjects of science and often into other areas altogether – for example, philosophy, religion, sociology or history. The key is to develop a balanced dialogue, in which people's ideas and observations based on experience interact with robust knowledge built from careful experiment and deduction.

I hope you have found at least some of the discussions captured in this book interesting and that some concepts in science have become a little clearer for you. If so, it may help you feel more confident about raising questions or engaging in conversations with a scientific angle. Some people in the discussion group featured in the book have been meeting together for upwards of 13 years. From time to time they are asked what it is that they get from participating; they speak with one voice about the benefits and limitations of this informal way of engaging with scientific ideas.

The bad news they report is that they don't become instantly brilliant at science. After all, they're not studying on a daily basis, nor are they going into detail about the theories or the mathematics that underpin them. Nor are they engaging in any practical laboratory work. However much they wish they could, they simply don't remember everything they've learned. But then what student ever does? And there's no doubt that grappling with strange and difficult concepts with which they made little progress at school can be hard on occasions. Things can seem disconnected at first.

The good news is that many participants feel a growing sense of enlightenment over time, as one thing after another becomes clearer. They start to see links between things and begin to understand some underlying concepts that are unifying and powerful. People say they feel much more confident to read and enter into conversation about scientific things, and come to see they have sometimes been bamboozled by people who only appear to be more knowledgeable than them. With this growing confidence, people in the groups have opened themselves up to opportunities to find out about science, often in a social way. Some have joined Astronomical Societies or signed up to an Open University module. Many have got together in small groups to visit museums, attend public lectures and participate in science fairs. Some find that getting pleasure out of science also helps them engage more with their children and their schoolwork. Others find themselves feeling better informed about particular political or social issues, such as climate change or gene therapy.

These stories may have made you think back about your own experience of learning science, perhaps years, maybe decades ago. It may have been largely forgotten or even blanked out. If so, I hope this book has helped reframe the way you see the subject, to take a fresh look as an adult. Perhaps it may stir up ideas about looking a bit further into science, in an informal sort of way. Even as an individual, there are many things you can do to pursue this gently. A first step could be to look in the popular science section of your local library or bookshop, taking care to browse carefully. It's probably advisable to avoid books that become too technical too quickly, which might be off-putting.

Modern communications also offer many other resources to draw upon. Television is full of interesting documentaries on scientific topics, complemented by well-produced podcasts that explain scientific ideas clearly. The internet has brought several good websites into existence, many of which address the general public and combine accessible language with good images, animations and videos. Again, care needs to be taken to find clear, intelligible ones. For more detailed suggestions about how to engage with these types of public resource see chapter 9 of my book *Getting to Grips with Science: a fresh approach for the curious* and the accompanying website https://gtgwithscience.com/websites.

It's also fun to take a more social approach by visiting museums or other places of scientific interest. Even if you live some distance from a large city, you may well find interesting places to visit with a scientific slant. Interesting science is carried out in so many hidden places – power stations, research institutes, universities and colleges, hospitals, water treatment works, fisheries, food testing and pollution testing labs, for example – and most scientists are only too pleased to arrange a visit and discuss what they do.

One way to make all this happen, to motivate oneself, might be to form a discussion group like those that have inspired this book. To do so would simply involve gathering together a small group of like-minded people and finding a part-time or retired science teacher willing to try out a novel approach. If my experience is any example, the teacher would need to be ready to abandon many of their normal teaching practices. They need to be ready to listen hard and to take their cue from the group, rather than a textbook or syllabus. They have to forget about being an all-knowing expert (or pretending to be) – most of the questions they'll be asked will not lie in their field. The teacher's role will be to make people feel secure in expressing themselves and to seek out the underlying concepts at play rather than detailed factual knowledge – that can be back-filled later from the internet or books. Where topics get really interesting, a group visit to a local researcher may be the best way forward.

The long-term survival of the groups I have been involved in is also the result of some very practical considerations. Monthly meetings in the early evening seem to work best. It's important to avoid any rules about, or even expectations of, attendance. Adults, however motivated, are often busy people with complex work and home lives. The structure needs to allow them to drop in when they can – some will be more regular than others. The key point is that members of the group need to feel in control of the planning, and indeed the content, of the discussion sessions. It also helps to find a congenial setting such as a wine bar, cafe or community space that is neither intimidating nor uninviting (and not too noisy either!).

Perhaps, with a fair wind and a little planning, you might find a way to take your interest in science one step further, however tentatively. If you do so, my job has been done.

Appendix: Atoms, Elements and Molecules

Some underlying concepts crop up time and again in discussion and throughout this book. Of these the most frequent are those that describe the make-up of substances. They occur whether we are talking about haemoglobin in blood, cochineal in food colouring or oxygen in the air. You may a have a rough idea of the meaning of words such as element, molecule and atom, but the precise distinctions may need clarifying. As Julie once put it memorably, when a discussion about the nervous system was in full swing: 'Andrew, before we go any further: what exactly is a molecule?' 'Is it different from an element?' queried Sarah, underlining the general sense of confusion. In this Appendix we attempt to clarify these differences in meaning, starting with what is closest to everyday life: the substances we hear about or read about on labels.

Substances and mixtures

Most things we are familiar with in everyday life are mixtures of different substances: cough syrup, jam or air, for example. Underneath this apparent variety lies an important unifying simplicity: every kind of 'stuff' (or, more properly, material) is made exclusively of atoms.

In *mixtures*, the blend of ingredients can be varied – as any maker of jam knows. There is a different percentage of sugar in different jams, for example. The word *substance*, on the other hand, is reserved in science for materials that have a fixed composition: the blend of ingredients doesn't vary. This means the atoms that make up a substance are always in the same proportion. Salt, for example, is a combination of sodium and chorine, always in an exact 1:1 ratio, and carbon dioxide always comprises two oxygen atoms for each carbon atom. In a substance, these atoms are chemically bonded to one another.

Compounds

A jar of strawberry jam contains pectin, a car battery contains hydrochloric acid, the air we breathe out contains carbon dioxide. These are all examples of compounds. They are made up (or compounded) of more than one kind of atom. Most substances are in fact compounds. However, a few substances are not compounded, but are made up of only one type of atom – a bar of gold, for example.

Elements

Elements are substances that are made of one kind of atom; carbon, mercury and hydrogen, for example, are all elements and as such cannot be broken down into simpler components. There are over 100 elements, all charted in the famous Periodic Table. Only 92 elements are found in nature, and these are the ingredients from which everything we can see is made, including our own bodies. There are also about 25 *synthetic elements*, artificially made in our nuclear reactors and in the laboratory, for example plutonium. Clearly elements differ dramatically one from another. Carbon is a hard solid, mercury a liquid metal and hydrogen an invisible gas. The difference in their properties lies in the make-up of the atoms of the various elements.

Atoms

The atom is the basic unit that distinguishes one element from another. The overall architecture of all atoms is the same: each is composed of a number of particles arranged in the same way. But the number of particles in the atoms of the various elements does differ. In all atoms there is a number of positively charged particles (called protons), all clustered together in a hard kernel called a nucleus. An equal number of negatively charged particles (called electrons) is arranged well away from the nucleus, and these particles are in rapid motion around it. A carbon atom, for example, has six protons and an equal number of electrons, an atom of oxygen has eight protons and electrons and hydrogen has only one of each. The equal number of positive and negative particles means that matter is normally uncharged, or neutral, overall. The nucleus also contains a number of uncharged particles called neutrons, clustered together with the protons.

Atoms are the smallest units of matter in everyday circumstances. However, in the form in which we normally encounter things, they are often aggregated into larger units. In metals and minerals, the atoms are held together by electrical attraction. In many other substances, including most biological ones, the atoms are bonded to one another in the form of molecules.

Molecules

Molecules are the smallest unit in which many substances normally exist. A molecule is a group of atoms held together by bonds; carbon dioxide (CO_2) and water (H_2O) are familiar examples. In general molecules contain atoms of different elements – such as carbon, hydrogen and oxygen – but some, like the oxygen molecule we breathe (O_2), consist simply of atoms of the same type. Molecules can be as small as just two atoms bonded together or may contain dozens or hundreds of atoms. In living systems some types of molecules (proteins and DNA, for example) run to many thousands of atoms.

The majority of substances we encounter in this book are made up of molecules. Foods, for example, contain proteins and carbohydrates, which are large molecules containing many atoms. Other bodily materials, such as neurotransmitters, amino acids and sugars are also molecules. Two examples of molecular structure are illustrated here (Fig. A.1), using models in which the rounded shapes represent atoms. White ones represent hydrogen atoms, red are oxygen, black are carbon and blue are nitrogen.

Ions

An ion is an atom or molecule that is electrically charged because of an imbalance in the number of charged particles of which it is made. It has either more or fewer electrons than protons, and may thus be negatively or positively charged by the gain or loss of one or more electrons. For example, an ion of the calcium atom has two fewer electrons and is denoted by Ca^{2+} or Ca^{++}; while an ion of the atom chlorine has one extra electron and is denoted by Cl^-.

A few substances mentioned in the book are not structured as discrete molecules, but as an extended array of ions. In common salt, sodium chloride, for example, the ions are held together, as in other minerals, by electrical attraction (Fig. A.2). The sodium is in the form of a positively charged ion (Na^+) and the chloride a negatively charged ion (Cl^-). The two types of ion are attracted to each other, holding the substance together.

(a) (b)

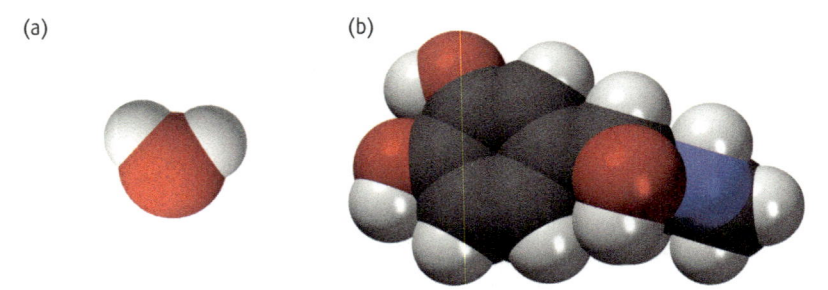

Fig. A.1 Models of molecules

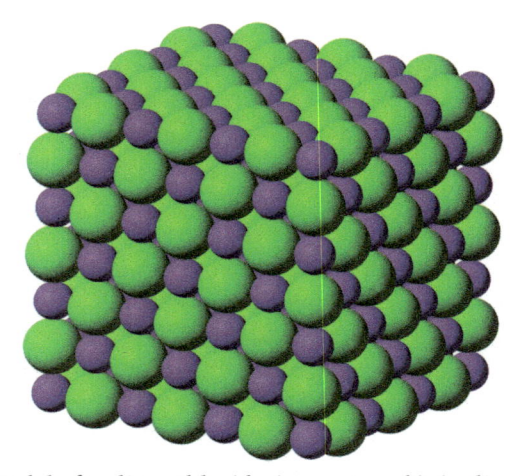

Fig. A.2 Model of sodium chloride (common salt) (sodium is violet, chlorine is green)

Further Resources

Fortunately a wide range of resources is available today to help people pursue their interest in science. A large number of books on specific topics have been written in an accessible fashion and websites, podcasts and online videos have been developed for the non-specialist. In addition museums, festivals, open days and public events provide live opportunities to find out more about science and the work of contemporary scientists.

Examples of such resources and their characteristics are given in a chapter on 'Taking Things Further' in my recent book *Getting to Grips with Science: A Fresh Approach for the Curious* (Imperial College Press, 2015). A website associated with that book also summarises this and provides links to some useful resources. On the website you are invited to share your experience of popular science resources, read the views of others and suggest ideas yourself. See: https://gtgwithscience.com.

Below are some ideas about the types of resource available:

Books

There are several kinds of book about science for the general reader. For example:

> *Specific subject areas.* Books written by scientists tend to focus, as you might expect, on their particular field. Popular subjects include evolution, genetics, cosmology and brain science.
>
> *History and philosophy of science.* The story of science is often written about by professional writers, ranging from Bill Bryson's *A Short History of Nearly Everything* to John Gribbin's *Science: a History*.
>
> *Biographies of scientists.* Great scientists themselves can be a source of fascination. Their lives and work are described in highly

readable biographies such as Georgina Ferry's *Dorothy Hodgkin: a Life* and Graham Farmelo's life of Paul Dirac, *The Strangest Man*.

The sheer number and range of readable science books available today makes it impossible to recommend individual ones. In addition, it's the views of the general reader rather than the science specialist that matter most in this case. For this reason you are invited to read and contribute to the comments of other readers on the website https://gtgwithscience.com/books. Help this list to grow by recommending your favourite titles and joining in the discussions they provoke.

Websites

Scientific societies such as The Royal Society, The Society of Biology, The Institute of Physics and The Royal Society of Chemistry have pages dedicated to the general reader. Many other organisations, for example the Women's Institute, the BBC and the Open University, have web pages linking science to everyday life, as do organisations set up specifically for this purpose, such as The Naked Scientists. Links to these organisations are available at https://gtgwithscience.com/websites.

Other websites offer science-based help for everyday issues. For example the NHS Choices website describes various kinds of condition and explains how treatments work. The BBC Bitesize website has useful information to help with GCSE science.

Audio and video

Recorded talks about science, both audio and video, can be found on the internet under, for example, the BBC or *TED Talks*. *Little Atoms* is an interesting source of podcasts. Live talks are offered by many local organisations, including universities, the University of the Third Age and Women's Institutes, as well as local societies for wildlife, geology and other science-based topics.

Museums and other venues

There are science and/or technology museums in a number of cities. Research institutes and organisations involved with science, including hospitals, water and food companies and power stations, often welcome visits.

Festivals and events

Festivals of science are organised by organisations such as The Royal Society and the British Science Association. Many universities, research institutes and 'public engagement' bodies organise open days, demonstrations and talks on scientific matters of interest to the public.

Index

absorption of light 19, 42
AC *see* alternating current
action at a distance 128
addiction to foods 9
adolescent
   behaviour 92, 175, 190
   brain 174
adrenaline 94
alkaloids 8
alternating current (AC) 165
alveoli 27
ambiguity
   in use of words 36
   in use of models 58
amps 157, 163
animal culture 184, 187
antibiotic 102–112
   resistance to 102
   molecules 108
   allergic response to 107
anti-viral drugs 112
apes 186–8
appetite hormones 11, 91
arborisation 174
Archimedes principle 117
atmosphere, Earth's 35, 43, 44, 47, 133
atom, structure of 59, 119, 160, 196
ATP (Adenosine Tri – Phosphate) 142, 153
autism 97
axon 67, 69

bacteria 102–112
beta blocker 97
bipolar disorder 176
bonds
   double and single 22
   hydrogen 119
brain 72–84, 171
   adolescent 174–5
   development 173
   neurons and synapses 77–8
   pruning 174
   right, left specialisation 74, 173
   role in seeing 69
   structure and function 75, 172
brain and hormones 82
brain and ageing 173
brain and language 77, 173, 180
brain and learning 80, 177–9

Calorie 138, 153
cell 93–4, 151
   membrane 27, 67, 141, 151–2
   signalling 93
centripetal force 124
cerebellum 172
cerebral cortex 76, 172
cerebrum 172
charge, electric 162
chemical groups 90
chlorophyll 21–2
circular motion 131–2
colour
   in food 9
   fading of 16, 23
   chemical basis of 18
   of blood 26, 29
compound 196
conduction of heat energy 143
conductivity, thermal 148
convection of heat energy 143
cooling, Newton's law of 143
culture, animal 184, 187
current
   electric 157, 161
   alternating 165
   ocean 126
curvature of space 133–5

deceit and the brain 179
density 115–6
depression 73, 80, 97, 174
diabetes 10, 97
discussion group 2, 3, 192, 193
dopamine 79, 176
drugs
   development of 96
   anti-viral 112
dyes 18

Earth, gravitational field of 122, 129
earth, electrical 167–8
Einstein 56–7
   gravitational theory of, 133–5
electric charge 162
electric current 157, 161
electric field 129
electricity 157–69
   in nerves 66
   mains 159, 165
   static 159

electromagnetic waves 43, 45
electron 160, 196
electronics 162
element 196
emotion 74, 179
energy 137–42
   chemical 140, 153
   conservation of 140
   definition 139
   heat 143–9
   in food 141
   levels in molecules 19, 23
   nuclear 34
evolution
   of taste 6
   of bacteria 107
   of schizophrenia 180
eye 64

field
   concept of 128
   electric 129
   magnetic 128
   gravitational 129
'fight or flight' response 95, 97
floating 115
food preferences 5–15
force
   centripetal 124
   gravitational 54, 123, 131–4
functional MRI *see* MRI
fuse 163

Galileo and gravity 44,128
gene
   defect 97, 111
   therapy 111
generator, electrical 165
ghrelin 11, 91
gland 91
   adrenal 95
   pituitary 82
glucose 10, 141, 153
gravitational force 54, 123, 131–4
gravity 122, 127, 129–32
   and curvature of space 127

H_2O molecule 118
haemoglobin 23, 28
heat energy, flow of 143–9
hormone 10, 82–4
   appetite 11, 91
   definition of 88–9
   replacement therapy 87
hormones and the brain 82
hot flush *see* menopause
hydrogen bond 119
hypothalamus 11, 83, 87

ice 115, 117
iceberg 114, 116
insulation, thermal 149
insulin 10, 97
intensity 41, 62
internal energy 147
inverse square law 42, 132

ion
   definition of 197
   and salty taste 8
   in nerves 66

language and the brain 77, 173, 180
laws of motion 128
learning and the brain 80, 177–9
leptin 11, 91
light
   absorption of 19–21
   as photons or waves 45, 57
   intensity of 41, 62
   models of 45
   ray 31, 36
   reflected 21, 29, 36–40, 46
   sources of 33
   speed of 44, 56
   white 20
lipid 151
liquid state 118
live wire 167
lock-and-key mechanism 94, 97

magnetic resonance 170
mains electricity 159, 165
mass 122, 129
membrane 27, 67, 141, 151–2
   of nerve cell 69
memory, role in learning 80, 178
menopause 86
mental illness 180
   see also autism, bipolar disorder,
      depression, schizophrenia
metabolism 11
metaphor, role of in science 27, 50, 58
mirror, reflection in 37
mitochondria 142, 153, 154
mixture 195
models 50–5
   of light 45
molecule, explanation of 197
Moon 36, 54
   and the tides 123–7
   gravitational field of 129–32
MRI 69, 76, 170–1, 177

neap and spring tides 126
nerve 66–9
nerve cell, in brain 77–80
neurogenesis 174
neuron *see* nerve cell
neurotransmitter 78, 176
neutral wire 167
neutron 196
Newton
   and gravity 54, 123, 128, 130
   law of cooling 143
   laws of motion 128
   theory of light 20, 57
nuclear fusion 34
nucleus of an atom 19, 34, 160, 171, 196
nucleus of a cell 100, 152, 154

oestrogen 86, 94
optic nerve 66, 69

optical illusions 62
orbital motion of the universe 131
oxygen
   and combustion 140
   and respiration 153
   in blood 26–8
oxytocin 96

Parkinson's disease 78
photolysis 24
pigment 17, 21–3
pituitary gland 82
planets, orbital motion of 131
plasma 33
power, electrical 158, 163
power station 165–8
primate 186–90
   behaviour 189
   personality 188
proton 160, 196
pruning, in the brain 174
psychopathy and the brain 181
puberty and the brain 175

quantum theory 19, 45, 49

radiation
   meaning of 41
   of heat energy 143
radioactivity 34
ray (of light) 31, 36
receptor cell 12, 94, 97, 107
reflection (of light) 36–40
relativity 56, 134
religion 187, 189
resistance, electrical 164
resistance, antibiotic 102
retina 65, 69
risk-taking 175, 190
   in the adolescent brain 175
   in cultures 190

scattering (of sunlight) 47
schizophrenia 174, 176, 180
seeing
   meaning of 60
   mechanism of 69
   psychology of 70
serotonin 80, 96

short- and long-sightedness 64
sky 46–7
social emotions 179
solid state 118
space 44, 46, 56, 134
   curvature of 133–5
spacetime 134
spectrum, electromagnetic 43
stars 33, 35, 42, 44, 101
static electricity 159
steroid 89
substance, meaning of 195
sugars 10
Sun 33–5, 47
   effect on tides 126–7
synapse 78, 173, 177
   density of, in the brain 174

taboos (in animal societies) 188
taste 7
   bud 8, 12
   receptor 8, 12
   psychology of 8
teenager *see* adolescent
teleology 106
temperature 144–8
'three-parent' baby 153
tides 121–6
   high and low 123
   neap and spring 126

ultraviolet radiation 24, 35, 42
universe, nature of 35, 55, 130, 135, 137

vacuum 119
virus 101, 108–11
visual cortex 70
voltage 158, 161
volume 116

water, anomalous behaviour of 117
watts 159
wave–particle duality 45
wavelength 42
waves
   electromagnetic 43, 45
   ocean 125
weight, meaning of 115, 116, 122, 129
work, scientific meaning of 139

Lightning Source UK Ltd.
Milton Keynes UK
UKOW07f1314180517
301484UK00004B/20/P